The MATLAB PROJECT BOOK

for LINEAR ALGEBRA

RICK L. SMITH
UNIVERSITY OF FLORIDA

 PRENTICE HALL Upper Saddle River, NJ 07458

Acquisition Editor: *George Lobell*
Assistant Editor: *Audra J. Walsh*
Production Editor: *Lorena Cerisano*
Special Projects Manager: *Barbara A. Murray*
Production Coordinator/Buyer: *Alan Fischer*
Supplement Cover Manager: *Paul Gourhan*

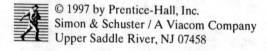

Printed in the United States of America

10 9 8 7 6 5 4 3 2 1

ISBN 0-13-521337-1

Prentice-Hall International (UK) Limited, *London*
Prentice-Hall of Australia Pty. Limited, *Sydney*
Prentice-Hall Canada, Inc., *Toronto*
Prentice-Hall Hispanoamericana, S.A., *Mexico*
Prentice-Hall of India Private Limited, *New Delhi*
Prentice-Hall of Japan, Inc., *Tokyo*
Simon & Schuster Asia Pte. Ltd., *Singapore*
Editora Prentice-Hall do Brasil, Ltda., *Rio de Janeiro*

To the memory of my mother
Isabelle Johnston Smith

Contents

Questions and Answers

This section is an introduction to the book by way of answering the most frequently asked questions.

Question 1. What is in this book?

There are 34 linear algebra projects in this book. The projects are intended to be worked independently by students on a computer. It is assumed that the computer is equipped with the software package MATLAB[r] and the instructor is sufficiently conversant with MATLAB to answer questions and grade these projects. There is a tutorial included to assist in learning MATLAB. The book is written so that the major theorems and definitions for a course in linear algebra have been included. This has been done in an attempt to make the book compatible with any linear algebra text. Those theorems which are standard fare for such a course are not proven; the others include proofs or proof sketches.

Question 2. What is MATLAB and why is this book designed for it?

MATLAB is numeric-graphic software which also includes a programming environment. It is marketed and supported by

The MathWorks, Inc.

Cochituate Place

24 Prime Park Way

Natick, MA 01760

The numerical functions are based on the LINPACK and EISPACK libraries which are among the most highly tested and reliable programs available for solving linear equations and computing eigenvalues. There is a symbolic toolkit available which is based on Maple[r] , one of the premier symbolic mathematics environments. MATLAB is easy to learn and easy to use. It runs on DOS, Windows, Macintosh, VAX, and Unix machines - to name a few - and programs written on one machine are easily ported to another. MATLAB is research grade software used by mathematicians, scientists, and engineers. Students who learn to use MATLAB will be able to use it in later courses. Instructors using MATLAB in their own work will be able to quickly move interesting examples from their research environment to the classroom.

Question 3. How do I (an instructor) use this book?

Here is how I use the book; you may choose to do things differently. Early in the semester I set aside one class period where we meet in the computer lab. In that one period I introduce the students to the operating system, the window system, and MATLAB. They start working through the *MATLAB Tutorial* at that time. Quickly, after that meeting I assign a project, which will be due in 10-14 days. I assign around six projects during the semester. The projects are collected instead of homework. I give weekly quizzes to keep the students current with the material of the course and to exercise them in hand calculation. Sometimes a project has more new material than I normally expect them to learn on their own. When that happens I will take a class period to lecture on the background material of the project. The vocabulary of the class is changed by the use of MATLAB. We have a precise formal language for expressing the algorithms of linear algebra. I expect them to master this language and I actually ask MATLAB questions on an exam. Here is a sample MATLAB question with acceptable solutions.

Write a MATLAB program b=invert(A) which returns b=1 if A is invertible and b=0 otherwise. Assuming A is a square matrix,
   Solution 1: b=1-(det(A)==0)
   Solution 2: [m,n]=size(A); b=(rank(A)==n)
   Solution 3: B=rref(A); b=prod(diag(B))
I am sure that you can think of other solutions. All of these solutions require some mastery of the theory and MATLAB.

Question 4. How do I (a student) use this book?

After being introduced to MATLAB, say, through the *MATLAB Tutorial*, you are ready to begin working projects. I suggest that before sitting down at a computer that you read the entire project including the assigned exercises to get a sense of what the project is about. Much of this first reading will be mysterious since there are MATLAB commands to be tried, but this should pique your interest in what MATLAB will do. Now go to the computer and begin your second reading this time trying the MATLAB commands. When you encounter a new or forgotten command, try MATLAB's help feature. It is possible that there is a reference to a function in another project. If this is the case, there is a page reference to the function in the *MATLAB Index*.

Question 5. What do the students turn in for grading?

This is where MATLAB's diary command is particularly useful. For those exercises which are interactive I have them turn in a diary of that exercise and for the programming exercises I have them turn in a copy of the program. I used to have them print this on paper, but now they mail me their answers electronically. Graphics plots still need to be printed.

Question 6. Why are there so many projects?

Having confessed in Question 3 that I actually assign only a fraction of the projects in this book, why have so many projects been included? There are several reasons. The instructor is given a choice of what to assign by having the extra projects. Some of the projects are included as reference, notably *Polynomials, Complex Numbers,* and *Graphics* and as such are rarely assigned. Primarily, I hope that some students, excited by mathematics and MATLAB, would work their way through the book.

Question 7. What course should this book be used in?

I use this book in an Elementary Linear Algebra course and most of the projects have been written with that audience in mind. I have also used it in a more advanced course on numerical methods. The projects work well with individual study courses for students who would like a mathematical introduction to computing and programming.

Question 8. Is this book a complete MATLAB reference?

No. I think the book is useful for learning MATLAB, linear algebra, and generally having fun with MATLAB, but no attempt has been made to catalog all of MATLAB's various and changing features. The easiest way to learn about the functions in your particular version of MATLAB is to use MATLAB's `help` feature.

Question 9. How can I find the MATLAB command that I need?

There are two indexes included in this book. There is a standard index of linear algebra terms and an alphabetical listing of the MATLAB commands which are used in the book. If the function that you need is not in the book, you can always use MATLAB's `help` command to look at the entire list of functions that you have available.

Question 10. Do the programs in this book work?

They used to. These programs have all been tested on MATLAB version 4.2. As MATLAB changes with different versions some of the commands and features used in the programs may be changed, causing the programs to fail. I have tried to write the programs as clearly and concisely as possible using what I think should be enduring MATLAB features. If you type in a program and it does not work, carefully check for spelling, punctuation, and misreadings such as "1" (one) for "l" (ell.)

Question 11. Do you need to program to use this book?

No. There are nonprogramming activities in the projects eliminating the need to program. Since programs have been listed in the book and those listed are usually referenced, it is necessary for someone to type these programs into the system in use. This is not programming, it is a typing exercise, but I regard it as a valid student activity though some instructors may choose to have these programs loaded into the system for the students. Having said that, let me emphasize that I always require my students to program in this course. Programming is an activity which requires abstract, analytic thinking. As such I regard programming as an appropriate pedagogical activity. Programming is rapidly becoming necessary for all mathematicians, scientists and engineers, and MATLAB is an excellent environment in which one can learn programming.

Question 12. Is this a text book?

It is certainly not a textbook in the sense of what is called a textbook these days. It was written to supplement a variety of textbooks, and as a supplement it omits many explanations about mathematical concepts while attempting to provide explanations about computational issues. Also omitted are the large number of rote exercises included in most texts. I do use this book for individual study courses.

Question 13. Did anyone help you write this?

Yes, many people. I would like to acknowledge the assistance and support of my colleagues Phil Bacon, Tom Bowman, Beverly Brechner, Bruce Edwards, Gérard Emch, Albert Fathi, Livio Flaminio, Joe Glover, Bill Hager, Chat Ho, Jonathon King, Panos Livadas, Bernard Mair, Jorge Martinez, Kermit Sigmon, Chris Stark, Arun Varma, and Neil White. Many students have contributed, in particular, Randy Fischer, Mike Jamieson, Tomas Horak, Jun Li, Bill Moser, and Scott Woodward. Most importantly, I thank my wife, Jane, and children, Erin, Lauren, Scott, and Mark, for their patience while this book was being extracted.

Basics: Vectors and Matrices

This introductory section contains the basic definitions and notation for the rest of the book.

VECTORS

The set of real numbers is denoted by R. The set of complex numbers, that is, those numbers of the form $a + bi$ where a and b are real numbers and $i = \sqrt{-1}$, is denoted by C. A *scalar* is either a real number or a complex number. A *vector* is a sequence $x = (x_1, x_2, \ldots, x_n)$ where each entry x_i is a scalar. The *zero vector* is $\vec{0} = (0, \ldots, 0)$. The *length* or *size* of the vector x is n = the number of entries in x. The set of vectors of length n with real entries is denoted R^n. The *magnitude* of x is

$$\|x\| = \sqrt{x_1^2 + x_2^2 + \ldots + x_n^2}.$$

There is some possible confusion of terminology here since some books will call $\|x\|$ the length. The magnitude is frequently called the *2-norm* or the *Euclidean norm.* For more about norms see *Norms and Condition Numbers.* Notice that $\|\vec{0}\| = 0$. We say that x is a *unit vector* if $\|x\| = 1$. The *dot product* of two vectors in R^n, $x = (x_1, \ldots, x_n)$ and $y = (y_1, \ldots, y_n)$, is

$$x \cdot y = \sum_{i=1}^{n} x_i y_i.$$

Notice that $x \cdot \vec{0} = 0$ and for all x, $\|x\|^2 = x \cdot x$. The geometric significance of the dot product goes back to the formula for vectors x and y of size 2 or 3

$$x \cdot y = \|x\| \|y\| \cos(\theta)$$

where θ is the angle between x and y. If $\theta = 90°$, then this gives $x \cdot y = 0$. Tradition now defines two nonzeros vectors x and y to be *orthogonal* or *perpendicular* if $x \cdot y = 0$. The sum of two vectors of the same size is

$$x + y = (x_1 + y_1, \ldots, x_n + y_n)$$

and if r is a scalar, then the scalar product is

$$rx = (rx_1, \ldots, rx_n).$$

From these operations we get the following basic properties.

Theorem 1. *For all vectors* $x, y, z \in R^n$ *and scalars* r,

    (1) $x \cdot y = y \cdot x$
    (2) $x \cdot (y + z) = x \cdot y + x \cdot z$
    (3) $r\,(x \cdot y) = (r\,x) \cdot y = x \cdot (r\,y)$
    (4) $x \cdot x > 0$ if $x \neq \vec{0}$
    (5) $\|r\,x\| = |r|\,\|x\|$

MATRICES

A *matrix* is a rectangular array of scalars in the following form:

$$A = \begin{bmatrix} a_{11} & \cdots & a_{1j} & \cdots & a_{1n} \\ \vdots & & \vdots & & \vdots \\ a_{i1} & \cdots & a_{ij} & \cdots & a_{in} \\ \vdots & & \vdots & & \vdots \\ a_{m1} & \cdots & a_{mj} & \cdots & a_{mn} \end{bmatrix}.$$

The subscripts have been carefully arranged; a_{ij} always represents the entry in the ith row and the jth column. In MATLAB, function notation is used, so that $A(i,j) = a_{ij}$. We say that A has *size* $m \times n$ (read "m by n" for m rows and n columns). An $n \times n$ matrix is called a *square* matrix.

In matrix notation a vector can be represented as either a row matrix ($1 \times n$ matrix) $[x_1 \ x_2 \ \ldots \ x_n]$ or a column matrix ($n \times 1$ matrix)

$$\begin{bmatrix} x_1 \\ x_2 \\ \vdots \\ x_n \end{bmatrix}$$

Suppose A is an $m \times n$ matrix with the (i,j) entry denoted a_{ij} and B is an $m \times n$ matrix with the (i,j) entry denoted b_{ij}. Then we can add according to the rule $a_{ij} + b_{ij}$ (add corresponding entries) to get $A + B$. Subtraction is defined by the rule $a_{ij} - b_{ij}$ to get $A - B$. For a scalar r we can multiply rA according to the rule ra_{ij} (multiply each entry of A by r).

Suppose that A is an $m \times n$ matrix and B is an $n \times k$ matrix then we can multiply $AB = C$ to get an $m \times k$ matrix according to the rule

$$c_{ij} = \sum_{l=1}^{n} a_{il}b_{lj} = (a_{i1}, \ldots, a_{in}) \cdot (b_{1j}, \ldots, b_{nj}).$$

The vector $(a_{i1}, \ldots, a_{in})$ is the ith row of A and $(b_{1j}, \ldots, b_{nj})$ is the jth column of B, which is the usual way we think about matrix multiplication. The notation A*B is used in MATLAB for matrix multiplication. If A is a $1 \times n$ row matrix and B is a $n \times 1$ column matrix, then AB is a 1×1 matrix whose entry is the dot product. If A is an $m \times n$ matrix and v is a $n \times 1$ matrix, then Av is a $m \times 1$ matrix. If $B = [v_1 \ldots v_k]$ where $v_1, \ldots, v_k$ are the columns of B, then $AB = [Av_1 \ldots Av_n]$.

It is important to note that the rule $AB = BA$ does not hold in general. One way to see this is to let A be a 100×1 matrix and B be a 1×100 matrix, then AB is 100×100, while BA is 1×1. It is also easy to check that $AB \neq BA$ for the square matrices

$$A = \begin{bmatrix} 1 & 1 \\ 0 & 0 \end{bmatrix} \text{ and } B = \begin{bmatrix} 1 & 0 \\ 1 & 0 \end{bmatrix}.$$

Many of the other properties that we expect do hold.

Theorem 2. *Assuming that A, B, and C have sizes for which the following equations make sense*

    (1) $A(B + C) = AB + AC$
    (2) $(B + C)A = BA + CA$
    (3) $(rA)B = A(rB) = r(AB)$

The **transpose** of the matrix A is the matrix A^T given by

$$A^T = \begin{bmatrix} a_{11} & \cdots & a_{i1} & \cdots & a_{m1} \\ \vdots & & \vdots & & \vdots \\ a_{1j} & \cdots & a_{ij} & \cdots & a_{mi} \\ \vdots & & \vdots & & \vdots \\ a_{1n} & \cdots & a_{in} & \cdots & a_{mn} \end{bmatrix}$$

The effect of this operation is to turn the first row of A into the first column of A^T, the second row of A into the second column of A^T, and so on. Thus, if A is $m \times n$, then A^T is $n \times m$. One can check that

$$(A + B)^T = A^T + B^T \text{ and } (A^T)^T = A$$

In MATLAB the transpose operation is `A'` for matrices with real entries. If `A` has complex entries `A'` will take the complex conjugate of the entries as well as the transpose of `A`.

Theorem 3. $(AB)^T = B^T A^T$

Suppose that A and B are two $n \times 1$ column matices, then thinking of A and B as vectors the dot product is

$$A \cdot B = A^T B = B^T A.$$

Similarly, if A and B are two $1 \times n$ row vectors, then thinking of A and B as vectors the dot product is

$$A \cdot B = A B^T = B A^T.$$

A matrix, A, is called **symmetric** if $A = A^T$, or equivalently $a_{ij} = a_{ji}$ for all i, j. Using Theorem 3, one can check that for any matrix B, if we let $A = B^T B$, then A is a symmetric matrix. We wrap up this section with some of the special matrices for linear algebra.

The $m \times n$ matrix with just zero entries is generically referred to as the **zero matrix**

$$O = \begin{bmatrix} 0 & \cdots & 0 \\ \vdots & & \vdots \\ 0 & \cdots & 0 \end{bmatrix}.$$

MATLAB uses the notation `zeros(m,n)` for the $m \times n$ zero matrix. Certainly, $A + O = A = O + A$.

The $n \times n$ **identity matrix** is

$$
I_n = \begin{bmatrix} 1 & 0 & \cdots & 0 \\ 0 & 1 & \ddots & \vdots \\ \vdots & \ddots & \ddots & 0 \\ 0 & \cdots & 0 & 1 \end{bmatrix}.
$$

MATLAB uses the notation $\texttt{eye(n)} = I_n$.

Let

$$
e_1 = \begin{bmatrix} 1 \\ 0 \\ \vdots \\ 0 \end{bmatrix}, \; e_2 = \begin{bmatrix} 0 \\ 1 \\ \vdots \\ 0 \end{bmatrix}, \ldots, e_m = \begin{bmatrix} 0 \\ \vdots \\ 0 \\ 1 \end{bmatrix}.
$$

Note that $I_m = [e_1, e_2, \ldots, e_m]$. These vectors are called the **standard basis.** Now if v is a vector of size m, then $I_m v = v$, and thus if $A = [v_1, \ldots, v_n]$, then $I_m A = [I_m v_1, \ldots, I_m v_n] = A$. Similarly, $A I_n = A$.

Let $\text{ele1}(n, i, j)$ be the matrix resulting from switching the ith and j^{th} rows of the identity matrix I_n. Let $\text{ele2}(n, r, i)$ be the result of multiplying the ith row of I_n by $r \neq 0$. Let $\text{ele3}(n, r, i, j)$ be the result of multiplying the ith row of I_n by $r \neq 0$ and adding it to the jth row of I_n. These are the **elementary matrices.**

Theorem 4. *Suppose A is an $m \times n$ matrix.*

(1) *$\text{ele1}(m, i, j)A$ is the matrix which results from interchanging the ith and j^{th} rows of A.*

(2) *$\text{ele2}(m, r, i)A$ is the matrix which results from multiplying the ith row of A by the scalar r.*

(3) *$\text{ele3}(m, r, i, j)\, A$ is the matrix which results from multiplying the ith row of A by r and adding it to the jth row of A.*

A matrix A is **invertible** if there is a matrix B such that $AB = BA = I_n$. The matrix B is called the inverse of A and is written A^{-1} In MATLAB the inverse is $\texttt{inv(A)}$. See *Inverses* for ways to compute A^{-1}. In view of Theorem 4, we see that the elementary matrices are invertible.

$\text{ele1}(n, i, j) * \text{ele1}(n, i, j) = I_n$.

Thus $\text{ele1}(n, i, j)$ is the inverse of $\text{ele1}(n, i, j)$.

$\text{ele2}(n, r, i) * \text{ele2}(n, 1/r, i) = I_n$.

So that $\text{ele2}(n, r, i)^{-1}$ is $\text{ele2}(n, 1/r, i)$.

$\text{ele3}(n, r, i, j) * \text{ele3}(n, -r, i, j) = I_n$.

And so the inverse of $\text{ele3}(n, r, i, j)$ is $\text{ele3}(n, -r, i, j)$.

MATLAB Tutorial

This is an interactive introduction to MATLAB. A sequence of commands has been provided for you to try. After each command you should type the Return or Enter key to execute the command. MATLAB runs on many different systems. You should consult with your instructor or the MATLAB documents on the method for getting into the MATLAB system. In the course of the tutorial if you get stuck on what a command means type

> help *command name*

and then try the command again. You should record the outcome of the command either in the space provided on the right side of the page or in a notebook.

1. BUILDING MATRICES

MATLAB has many types of matrices which are built into the system. A 7×7 matrix with random entries is found by typing the commands followed by a return:

> rand(7)

> rand(2,5)

> help rand

Another special matrix, called a Hilbert matrix is

> hilb(5)

> help hilb

A 5×5 magic square is given by

> magic(5)

> help magic

Here are some of the standard matrices from linear algebra.

> eye(6)

> zeros(4,7)

> ones(5)

You can also build matrices of your own with any entries that you may want.

> [1 2 3 5 7 9]

> [1, 2, 3;4, 5, 6;7, 8, 9]

> [1 2 <RETURN> 3 4 <RETURN> 5 6]

Here <RETURN> means to type the Return or Enter key.

> [eye(2);zeros(2)]

> [eye(2);zeros(3)]

You should have got an error message on this. Why?

> [eye(2),ones(2,3)]

2. VARIABLES

MATLAB has built-in variables like pi, eps, and ans. You can learn their values.

> pi

> eps

> help eps

```
ans
```

At any time if you want to know the active variables try:

```
who

help who
```

The variable ans will keep track of the last output which was not assigned to another variable.

```
magic(6)

ans

x=ans

x

x=[x,ones(6)]

x
```

Since you have created a new variable, x, it should appear as an active variable.

```
who
```

To remove a variable try this.

```
clear x

x

who
```

3. FUNCTIONS

In *Basics: Vectors and Matrices*, some of the MATLAB functions were introduced. Here we will get some practice with those functions.

```
a=magic(4)
```

Take the transpose of a.

```
a'
```

If a has complex entries a' is not the transpose, rather it is the transpose of the conjugate of a. See *Complex Numbers* for practice with complex numbers.

```
3*a

-a

a+(-a)

b=max(a)

max(b)
```

Some functions can return more than one value. In the case of max it will return the maximum value and also the column index where the maximum value occurs.

```
[m,i]=max(b)

min(a)

b=2*ones(a)

a*b

a
```

Usually a dot in front of an operation will change the operation. In the case of multiplication, a.*b will make it into an entry-by-entry multiplication instead of the usual matrix multiplication.

```
a.*b (there is a dot there!)
x=5
x^2
a*a
a^2
a.^2 (another dot)
a
triu(a)
tril(a)
diag(a)
diag(diag(a))
c=rand(4,5)
size(c)
[m,n]=size(c)
m
d=.5-c
```

There are a many functions which we apply to scalars that can be applied to matrices also.

```
sin(d)
exp(d)
log(d)
abs(d)
```

MATLAB has functions to round floating point numbers to integers. These are round, fix, ceil, and floor. The next few commands will help you determine how these work.

```
f=[-.5  .1  .5]
round(f)
fix(f)
ceil(f)
floor(f)
sum(f)
prod(f)
```

4. RELATIONS AND LOGICAL OPERATIONS

In this section you should think of 1 as "true" and 0 as "false." The notations &, |, $\sim$ stand for "and," "or," and "not," respectively. The notation == is a check for equality.

```
a=[1,  0,  1,  0]
b=[1,  1,  0,  0]
a==b
a<=b
~a
a&b
a&~a
a|b
a|~a
```

There is a function , any , to determine if there is a nonzero entry in a vector, and a function , all , to determine if all the entries are nonzero.

```
a
any(a)
all(a)
c=zeros(1,4)
d=ones(1,4)
any(c)
all(d)
e=[a',b',c',d']
any(e)
all(e)
any(all(e))
```

5. COLON NOTATION

MATLAB also offers some very powerful ways for creating vectors.

```
x=-2:1
length(x)
-2:.5:1
-2:.2:1
a=magic(5)
a(2,3)
```

Now we will use the colon notation to get a column of a.

```
a(2,:)
a(:,3)
a
```

```
a(2:4,:)
a(:,3:5)
a(2:4,3:5)
a(1:2:5,:)
```

You can put a vector in the row or column position of a.

```
a(:,[1,  2,  5])
a([2,  5],[2,  4,  5])
```

You can also make assignment statements using a vector or a matrix.

```
b=magic(5)
b([1  2],:)=a([2  1],:)
a(:,[1  2])=b(:,[3  5])
a(:,[1  5])=a(:,[5  1])
a=a(:,5:-1:1)
```

When you insert a 0-1 vector into the column position, the columns which correspond to the 1's are displayed.

```
v=[0 1 0 1 1]
a(:,v)
a(v,:)
```

This has been a sample of the basic MATLAB functions and the matrix manipulation techniques. At the end of this tutorial there is a listing of some of MATLAB's functions. The functions that you have available will vary slightly from version to version of MATLAB. By typing help you can access a listing of all functions in the version that you are using.

6. MISCELLANEOUS FEATURES

You may have discovered by now that MATLAB is case sensitive, that is "a" is not the same as "A."

The MATLAB display only shows 5 digits in the default mode. The fact is that MATLAB always keeps and computes in a double precision 16 decimal places and rounds the display to 4 digits. The command

```
format long
```

will switch to display all 16 digits and

```
format short
```

will return to the shorter display. It is also possible to toggle back and forth in the floating point format display with the commands

```
format short e
```

and

```
format  long  e
```

It is not always necessary for MATLAB to display the results of a command to the screen. If you do not want the matrix A displayed, type A; to suppress it. When MATLAB is ready to proceed, the prompt >> will appear. Try this on a matrix.

Occasionally you will have spent much time creating matrices in the course of your MATLAB session and you would like to use these same matrices in your next session. You can save these values in a file by typing

    `save` *filename*

This creates a file *filename.mat* which contains the values of the variables from your session. If you do not want to save all variables there are two options. One is to clear the variables off with the command

    `clear a b c`

which will remove the variables `a,b,c`. The other option is to use the command

    `save` *filename* `x y z`

which will save the variables `x,y,z` in the file *filename.mat*. The variables can be reloaded in a future session by typing

    `load` *filename*

When you are ready to print out the results of a session, you can store the results in a file and print the file from the operating system using the "print" command appropriate for your operating system. The file is created using the command

    `diary` *filename*

Once a file name has been established you can toggle the diary with the commands

    `diary on`

and

    `diary off`

The command `diary` *filename* will copy anything which goes to the screen (other than graphics) to the specified file, *filename*. Since this is an ordinary ASCII file, you can edit it later. Discussion of print out for graphics is deferred to the project *Graphics* where MATLAB's graphics commands are presented.

7. PROGRAMMING IN MATLAB

MATLAB is also a programming language. By creating a file with the extension *.m* you can easily write and run programs. If you were to create a program file *myfile.m* in the MATLAB language, then you can make the command *myfile* from MATLAB and it will run like any other MATLAB function. You do not need to compile the program since MATLAB is an interpretative (not compiled) language. Such a file is called an *m-file*. I am going to describe the basic programming constructions. While there are other constructions available, if you master these you will be able to write clear programs.

I. Assignment

Assignment is the method of giving a value to a variable. You have already seen this in the interactive mode. We write `x=a` to give the value of `a` to `x`. Here is a short program illustrating the use of assignment.

```
function r=mod(a,d)

% r=mod(a,d). If a and d are integers,
% then r is the integer remainder of a
% after division by d. If a and d are
% integer matrices, then r is the matrix
% of remainders after division by
% corresponding  entries. Compare with
% MATLAB's rem.

r=a-d.*floor(a./d);
```

You should make a file named *mod.m* and enter this program exactly as it is written. Now assign some integer values for a and d. Run

```
mod(a,d)
```

This should run just like any built-in MATLAB function. Type

```
help mod
```

This should produce the seven lines of comments which follow the % signs. The % signs generally indicate that what follows on that line is a comment which will be ignored when the program is being executed. MATLAB will print to the screen the comments which follow the "function" declaration at the top of the file when the help command is used. In this way you can contribute to the help facility provided by MATLAB to quickly determine the behavior of the function. Type

```
type  mod
```

This will list out the entire file for your perusal. What does this one line program mean? The first line is the "function declaration." A file with the function declaration at the top is called a *function file*. In it the name of the function (which is always the same as the name of the file without the extension *.m*), the input variables (in this case a and d), and the output variables (in this case r) are declared. Next come the "help comments" which we have already discussed. Finally, we come to the meat of the program. The variable r is being assigned the value a-d.*floor(a./d); The operations on the right hand side of the assignment have the meaning which you have just been practicing (the / is division) with the "." meaning the entry-wise operation as opposed to a matrix operation. Finally, the ";" prevents printing the answer to the screen before the end of execution. You might try replacing the ";" with "," and running the program again just to see the difference.

II. Branching

Branching is the construction

if *condition, program* end

The *condition* is a MATLAB function and *program* is a program segment. The entire construction executes the *program* just in case the value of *condition* is not 0. If that value is 0, the control moves on to the next program construction. You should keep in mind that MATLAB regards a==b and a<=b as functions with values 0 or 1. Frequently, this construction is elaborated with

```
if  condition ,
     program1
else
     program2
end
```

In this case if *condition* is 0, then *program2* is executed. Another variation is

```
if condition1,
     program1
elseif condition2,
     program2
end
```

Now if *condition1* is not 0, then *program1* is executed. If *condition1* is 0 and if *condition2* is not 0, then *program2* is executed. Otherwise control is passed on to the next construction. Here is a short program to illustrate branching.

```
function b=even(n)

% b=even(n). If n is an even integer,
% then b=1, otherwise b=0.

if mod(n,2)==0,
    b=1;
    else b=0;
end
```

III. For Loops

A "for loop" is a construction of the form

```
for i=1:n,  program, end
```

Here we will repeat *program* once for each index value i. Here are some sample programs. The first is matrix addition.

```
function c=add(a,b)

% c=add(a,b). This is the function
% which adds the matrices a and b.
% It duplicates the MATLAB function
% a+b.

[m,n]=size(a);
[k,l]=size(b);
if m~=k | n~=l,
  error('matrices not the same size');
  return,
end
c=zeros(m,n);
for i=1:m,
   for j=1:n,
      c(i,j)=a(i,j)+b(i,j);
   end
end
```

The next program is matrix multiplication.

```
function c=mult(a,b)

% c=mult(a,b). This is the matrix product
% of the matrices a and b. It duplicates
% the MATLAB function c=a*b.

[m,n]=size(a);
[k,l]=size(b);
if n~=k,
   error('matrices are not compatible');
   return,
end,
c=zeros(m,l);
for i=1:m,
   for j=1:l,
      for p=1:n,
         c(i,j)=c(i,j)+a(i,p)*b(p,j);
      end
   end
end
```

For both of these programs you should notice the branch construction which follows the size statements. This is included as an error message. In the case of add, an error is made if we attempt to add matrices of different sizes, and in the case of mult it is an error to multiply if the matrix on the left does not have the same number of columns as the number of rows of the matrix on the right. Had these messages not been included and the error was made, MATLAB would have delivered another error message saying that the index exceeds the matrix dimensions. You will notice in the error message the use of single quotes. The words surrounded by the quotes will be treated as text and sent to the screen as the value of the variable c. Following the message is the command return, which is the directive to send the control back to the function which called add or return to the prompt. I usually only recommend using the return command in the context of an error message. Most MATLAB implementations have an error message function, either errmsg or error, which you might prefer to use. In the construction

```
for i=1:n, program, end
```

the index i may (in fact usually does) occur in some essential way inside *program*. MATLAB will allow you to put any vector in place of the vector 1:n in this construction. Thus the construction

```
for i=[2,4,5,6,10], program, end
```

is perfectly legitimate. In this case *program* will execute 5 times and the values for the variable i during execution are 2,4,5,6,10, in that order. The MATLAB developers went one step further. If you can put a vector in, why not put a matrix in? So, for example,

```
for i=magic(7), program, end
```

is also legal. Now *program* will execute 7 (=number of columns) times, and the values of i used in *program* will be successively the columns of magic(7).

IV. While Loops

A "while loop" is a construction of the form

```
while condition, program, end
```

where *condition* is a MATLAB function, as with the branching construction. The program *program* will execute successively as long as the value of *condition* is not 0. While loops carry an implicit danger in that there is no guarantee in general that you will exit a while loop. Here is a sample program using a while loop.

```
function l=twolog(n)

% l=twolog(n). l is the floor of
% the base 2 logarithm of n.

l=0;
m=2;
while m<=n
    l=l+1;
    m=2*m;
end
```

V. Recursion

Recursion is a devious construction which allows a function to call itself. Here is a simple example of recursion

```
function y=twoexp(n)

% y=twoexp(n). This is a recursive program
% for computing y=2^n. The program halts
% only if n is a nonnegative integer.

if n==0,
    y=1;
else
    y=2*twoexp(n-1);
end
```

The program has a branching construction built in. Many recursive programs do. The condition n==0 is the base of the recursion. This is the only way to get the program to stop calling itself. The "else" part is the recursion. Notice how the twoexp(n-1) occurs right there in the program which is defining twoexp(n)! The secret is that it is calling a lower value, n-1, and it will continue to do so until it gets down to n=0. A successful recursion is calling a lower value.

There are several dangers in using recursion. The first is that, like while loops, it is possible for the function to call itself forever and never return an answer. The second is that recursion can lead to redundant calculations which, though they may terminate, can be time consuming. The third danger is that while a recursive program is running it needs extra space to accomodate the overhead of the recursion. In numerical calculations on very large systems of equations memory space is frequently at a premium, and it should not be wasted on program overhead. With all of these bad possibilities why use recursion? It is not always bad; only in the hands of an inexperienced user. Recursive programs can be easier to write and read than nonrecursive programs. Some of the future projects illustrate good and poor uses of recursion.

VI. Miscellaneous Programming Items

It is possible to place a matrix valued function as the condition of a branching construction or a while loop. Thus the condition might be a matrix like ones(2) or eye(2). How would a construction like

```
if condition,
      program1,
else
      program2,
end
```

behave if *condition* is eye(2)? If all of the entries of *condition* are nonzero, *program1* will execute. So if *condition* is eye(2), *program2* will execute. If *condition* is ones(2), *program1* will execute since all of the entries are nonzero. A problematic construction occurs when you have

```
if A~=B, program, end.
```

You would like *program* to execute if the matrices A and B differ on some entry. Under the convention, *program* will only execute when they differ on *all* entries. There are various ways around this. One is the construction

```
if A==B, else program, end
```

which will pass control to the "else" part if A and B differ on at least one entry. Another is to convert A==B into a binary valued function by using all(all(A==B)). The inside all creates a binary vector whose i^{th} entry is 1 only if the ith column of A is the same as the ith column of B. The outside all produces a 1 if all the entries of the vector are 1. Thus if A and B differ on at least one entry, then all(all(A==B))=0. The construction

```
if ~all(all(A==B)), program, end
```

then behaves in the desired way.

Essentially, the same convention holds for the while construction.

```
while condition, program, end.
```

The program *program* will execute successively as long as every entry in *condition* is not 0, and the control passes out of the loop when at least one entry of *condition* is 0. Another problem occurs when you have a conjunction of conditions, as in

```
if condition1 & condition2, program, end
```

Of course, *program* will execute if both *condition1* and *condition2* are not zero. Suppose that *condition1* is 0 and *condition2* causes an error message. This might happen for i<=m & A(i,j)==0 where m is the number of columns of A. If i>m, then you would like to pass the control, but since A(i,j) makes no sense if i>m an error message will be dished up. Here you can nest the conditions.

```
if i<=m,
      if A(i,j)==0,
          program
      end
end
```

VII. Scripts

We have seen m-files which have the function declaration at the top. In practice these files create new MATLAB functions. In creating a function which we will call `fun(a)` we might use a variable like `x`. Now suppose the the variable `x` has a value in your session. What happens to the value of `x` after you make a call to `fun(a)`? Nothing. The only way to change the value of `x` when running `fun` is to assign `x=fun(a)`. The `x` inside the program `fun.m` behaves independently from the variable `x` in your session. This makes function files very natural to use. You do not need to worry about whether you have used a variable inside of a program which already has a value in your session.

A *script* is an m-file without the function declaration at the top. A script treats variables differently than a function file. In a script, if `x` appears in a program, which we will call `scrpt`, and `x` has a value in your session, then a call to `scrpt` might change the value of `x`. If you do not make that function declaration, then the variables in your session can be altered. Sometimes this is useful, but I recommend that you use function files.

VIII. Clearing

In the interactive portion of the tutorial you saw how to use the `clear` function to remove a variable from your session. It is also possible to use `clear` to remove a function from your session. When a function is "read" by MATLAB into the machine it resides in the machine's memory. If you find that you are short of memory it is possible to clear the function with the command `clear functions`. Occasionally, when you are in the edit-run cycle, the machine will not acknowledge changes which have been made in the function you are editing. When you suspect this is the case try clearing the function before running again.

IX. Suggestions

These are a few pointers about programming and programming in MATLAB in particular.

1) You should use the indented style that you have seen in the above programs. It makes the programs easier to read, the program syntax is easier to check, and it encourages you to think in terms of building your programs in blocks.

2) Put lots of comments in your program to tell the reader in plain English what is going on. Some day that reader will be you, and you will wonder what you did.

3) Put error messages in your programs like the ones above. As you go through this manual, your programs will build on each other. Error messages will help you debug future programs.

4) Always structure your output as if it will be the input of another function. For example, if your program has "yes-no" type output, do not have it return the words "yes" and "no," rather return 1 or 0, so that it can be used as a condition for a branch or while loop construction in the future.

5) In MATLAB, try to avoid loops in your programs. MATLAB is optimized to run the built-in functions. For a comparison, see how much faster `A*B` is over `mult(A,B)`. You will be amazed at how much economy can be achieved with MATLAB functions.

6) If you are having trouble writing a program, try to get a small part of it running and build on that.

A Short List of Key Words and Symbols

This is a starter list of words and symbols for MATLAB. For the complete list as well as any m-files which are included in your implementation of MATLAB type `help`. More information is in the *MATLAB User's Guide* by The MathWorks, Inc. and published by Prentice-Hall, Inc. There is a *MATLAB Index* on page 238 for finding functions used in this book. The *MATLAB Primer* by Kermit Sigmon published by CRC Press, Inc. is also a useful reference guide to MATLAB.

magic	round	help	function	==
eye	floor	format short	if	~=
diag	abs	format long	else	<
ones	sqrt	who	elseif	<=
zeros	sin	clear	end	>
rand	cos	exit	while	>=
hilb	tan	ans	return	&
tril	asin	load	=	|
triu	acos	save	;	~
sum	atan	diary	:	any
prod	exp	type	%	all
max	log	dir	,	+
min	eps			-
size	pi			*
length	sign			.*
				,
				/
				./
				^
				.^

Systems of Linear Equations

ABSTRACT
An introduction to linear systems and methods for finding solutions.

MATLAB COMMANDS
```
rref, sum, *, +, -, /
```

LINEAR ALGEBRA CONCEPTS
Elementary Row Operations

BACKGROUND

A *linear equation* is an equation of the form

$$a_1 x_1 + a_2 x_2 + \cdots + a_n x_n = b$$

where $a_1, \ldots, a_n, b$ are real or complex numbers and $x_1, \ldots, x_n$ are the variables. A graph of a linear equation with real coefficients in two variables is a line; in three variables a plane.

A *system of m linear equations in n variables* is written in the following form:

$$a_{11} x_1 + \ldots + a_{1j} x_j + \ldots + a_{1n} x_n = b_1$$

$$\vdots \qquad\qquad \vdots$$

$$a_{i1} x_1 + \ldots + a_{ij} x_j + \ldots + a_{in} x_n = b_i$$

$$\vdots \qquad\qquad \vdots$$

$$a_{m1} x_1 + \ldots + a_{mj} x_j + \ldots + a_{mn} x_n = b_m$$

This is called an *m × n system.* The subscripts are carefully arranged so that a_{ij} is the coefficient of the variable x_j in the ith equation. We are interested in the solutions which are common to all of the equations. In the case of three variables, where each equation represents a plane, the solutions graph as the intersection of planes, thus the solutions will form either a plane, a line, a single point, or the empty set. If all the $a_{i,j}$'s and b_i's are 0, the set of solutions will be all of R^3. We will typically represent such a system in a matrix equation. The matrix given below is called the *coefficient matrix* of the system.

$$A = \begin{bmatrix} a_{11} & \cdots & a_{1j} & \cdots & a_{1n} \\ \vdots & & \vdots & & \vdots \\ a_{i1} & \cdots & a_{ij} & \cdots & a_{in} \\ \vdots & & \vdots & & \vdots \\ a_{m1} & \cdots & a_{mj} & \cdots & a_{mn} \end{bmatrix}$$

If we let

$$x = \begin{bmatrix} x_1 \\ \vdots \\ x_j \\ \vdots \\ x_n \end{bmatrix} \text{ and } b = \begin{bmatrix} b_1 \\ \vdots \\ b_i \\ \vdots \\ b_m \end{bmatrix}$$

then by the definition of matrix multiplication the solutions to the matrix equation $Ax = b$ are the same as the solutions to the original system of equations. We will now consider how to solve $Ax = b$.

Let A be any matrix. An ***elementary row operation of type 1*** switches two rows of A. We will denote the operation which switches rows i and j, elerow1(A, i, j). In MATLAB try

```
A=magic(8);
A([1,4],:)=A([4,1],:)
```

An ***elementary row operation of type 2*** multiplies a row of A by a nonzero scalar r. We will denote the operation which multiplies the ith row of A by r, elerow2(A, r, i). In MATLAB this is done easily.

```
A(3,:)=10*A(3,:)
```

An ***elementary row operation of type 3*** multiplies a row of A by a scalar and adds that row to another row. We call the operation which multiplies the ith row of A by r and adds it to the jth row, elerow3(A, r, i, j). In MATLAB

```
A(8,:)=2*A(1,:)+A(8,:)
```

Returning to the problem of solving a system of equations. Starting with $Ax = b$, we form the ***augmented matrix***, $[A, b]$. $[A, b]$ is the matrix obtained by placing the column vector b at the far right of the matrix A. The strategy is to systematically apply elementary row operations to simplify $[A, b]$. The next theorem states the legitimacy of this strategy.

Theorem. *Let $Ax = b$ be a matrix equation and suppose that $[C, d]$ is the matrix obtained after applying some elementary row operations to the augmented matrix $[A, b]$. Then $Ax = b$ has the same solutions as $Cx = d$.*

The goal of applying the elementary row operations is to bring the matrix into a form where the solutions can be found easily. The form we seek is called reduced row echelon form. A matrix is in ***reduced row echelon form*** if

   (1)   The first nonzero entry in each row is a 1. This entry is called a ***leading*** 1 .

   (2)   The entries above and below in a column where a leading 1 occurs are 0.

   (3)   The leading 1 in the ith row is to the left of the leading 1 in the $i + 1^{st}$ row (the ith row is above the $i + 1^{st}$ row).

   (4)   Any rows which have only 0 entries are at the bottom of the matrix.

The MATLAB function `rref` uses the elementary row operations to convert a matrix into a reduced row echelon form matrix. You can use this function to see some matrices in reduced row echelon form try

```
rref(magic(4))
```

```
rref(magic(8))
```

```
rref(rand(5))
```

Suppose we start with the system

$$2x - 10y + 12z + 3w = 0$$
$$3x - 15y + 18z + 4w = 0$$
$$x - 4y + 5z + 13w = 0$$

We are going to go through this step by step so that you can get some idea of how `rref` works. We will use elementary row operations of type 2 to create leading 1's and then we use elementary row operations of type 3 to zero out the the entries above and below the leading 1's. The entry which is selected to become the leading 1 is called the *pivot*. Begin by creating the coefficient matrix:

```
A=[2, -10, 12, 3; 3,-15,18,4;1,-4,5, 13]
```

We see that the (1,1) entry is nonzero, so it will be our pivot.

```
A(1,:)=1/A(1,1)*A(1,:)
```

We now have a leading 1, so we use it to zero out the other entries in the first column.

```
A(2,:)=-A(2,1)*A(1,:)+A(2,:)
```

```
A(3,:)=-A(3,1)*A(1,:)+A(3,:)
```

We have created our first leading 1 and zeroed out below. Now move to the next row and the next column. We see that there is a 0 in the (2,2) position so we cannot use an elementary row operation of type 2 to turn it into a leading 1. There is a nonzero entry below it in the (3,2) position, so the (3,2) element will be the pivot. Use an elementary row operation of type 1 to move that nonzero entry into the (2,2) position, then wipe out above and below.

```
A([2,3],:)=A([3,2],:)
```

```
A(1,:)=-A(1,2)*A(2,:)+A(1,:)
```

We have a leading 1 in the second row. Move to the third row and the third column. In the (3,3) position there is a zero, but this time there is no nonzero entry below. So we move on to the (3,4) position. There is a nonzero entry. It becomes the pivot. Turn it into a leading 1 and zero out above.

```
A(3,:)=1/A(3,4)*A(3,:)
```

```
A(2,:)=-A(2,4)*A(3,:)+A(2,:)
```

```
A(1,:)=-A(1,4)*A(3,:)+A(1,:)
```

We are now in reduced row echelon form. If you want to see a program for placing a matrix in reduced row echleon form, type in MATLAB

```
type rref
```

What are the advantages of having a reduced row echelon form? Return to the example just worked above and write out the system of equations which go with the reduced row echelon form.

$$x + z = 0$$
$$y - z = 0$$
$$w = 0$$

The variables x, y, w correspond to the leading ones. We will call these **leading variables.** The variable z which is not a leading variable is called a *free variable* or a *parameter.* We can always solve for the leading variables in terms of the free variables.

$$x = -z$$
$$y = z$$
$$w = 0$$

You can let z take any value and the values for the leading variables will be determined. Thus letting $z = 0$ gives the solution $x = 0$, $y = 0$, $w = 0$, or letting $z = 1$ gives $x = -1$, $y = 1$, $w = 0$. And so on for any value of z you might choose.

In MATLAB we can generate these solutions easily using an auxiliary vector L which tells us the columns where the leading ones occur. Here is the short program for creating L. Create an m-file called lead.m and enter the following program.

```
function L=lead(A)
% L=lead(A). The vector L has the values
% L(i)=1 if the i-th column of the reduced
% row echelon form contains a leading 1,
% and 0 otherwise.
[m,n]=size(A);
R=rref(A);
L=zeros(1,n);
row=1; col=1;
while col<=n & row<=m,
    if R(row,col)==1,
        L(col)=1; row=row+1;
    end
    col=col+1;
end
```

At this stage your A should be in reduced row echelon form. If not let A=rref(A). In the next few exercises we will assume that you have L=lead(A). You can assign L=[1 1 0 1] for the purposes of these exercises.

```
L=lead(A)
```

```
F=~L
```

Notice that F indicates the columns where the free variables occur. The number of leading variables is given by

```
r=sum(L)
```

and the number of free variables is given by

```
s=sum(F)
```

Should there be any rows of zeros at the bottom of A we need to remove them with

```
A=A(1:r,:)
```

We now start to make a solution vector. First create a vector of zeros.

```
x=zeros(size(L))';
```

If you missed taking the transpose, do it now with x=x' so that your x should be a column vector. We will assign the free variables random values.

```
x(F)=rand(s,1)
```

We need to assign values to the leading variables. Look at

```
A*x
```

This has multiplied the free variables by the appropriate entries of A and the leading ones of A have been multiplied by 0's. To solve for the leading variables, let

```
x(L)=-A*x
```

Check this quickly with

```
A*x
```

Why does this work? Suppose for simplicity that we can write

$$x = \begin{bmatrix} x(L) \\ x(F) \end{bmatrix} = \begin{bmatrix} x(L) \\ \vec{0} \end{bmatrix} + \begin{bmatrix} \vec{0} \\ x(F) \end{bmatrix}.$$

Since $A(:, L) = I_r$, the $r \times r$ identity, we see that

$$A \begin{bmatrix} x(L) \\ \vec{0} \end{bmatrix} = x(L) = -A \begin{bmatrix} \vec{0} \\ x(F) \end{bmatrix}$$

and thus

$$Ax = A \begin{bmatrix} x(L) \\ \vec{0} \end{bmatrix} + A \begin{bmatrix} \vec{0} \\ x(F) \end{bmatrix} = x(L) - x(L) = \vec{0}.$$

The system we have been working with has the form $Ax = b$ in matrix notation. When $b = \vec{0}$, as we have in this example, we say the system is **homogeneous.** Let's modify the system by changing b so that the system is no longer homogeneous. Rebuild A to be

```
A=[2,-10,12,3;3,-15,18,4;1,-4,5,13]
```

Make the following augmented matrix:

```
b=(1:3)', E=[A,b]
R=rref(E)
```

We want to find solutions to `Ax=b`. When we return to the system of equations from the reduced row echelon form we get

$$x + z = 72$$
$$y - z = 14$$
$$w = -1$$

Solving for the leading variables in terms of the free variables we get

$$x = 72 - z$$
$$y = 14 + z$$
$$w = -1$$

and again we see that the leading variables are determined by the free variables.

```
C=R(:,1:4)
d=R(:,5)
```

We know from Theorem 1 that `Ax=b` has the same solutions as `Cx=d`. Use MATLAB to create a solution. Watch the transpose!

```
L=lead(A);
x=zeros(size(L))'
x(F)=rand(s,1)
```

The difference now is the presence of d but this requires only a minor modification.

```
x(L)=d-C*x
```

Every homogeneous system $Ax = \vec{0}$ has a solution given by $x = \vec{0}$. This is not true when $b \neq \vec{0}$. If we modify the matrix C slightly

```
C(3,4)=0
```

The new system has the equations

$$x + z = 72$$
$$y - z = 14$$
$$0 = -1$$

The last equation cannot be satisfied by any choice of values for the variables and so the system has no solution. When a system has no solution we say that it is ***inconsistent,*** and when there is a solution it is ***consistent.*** Using `lead` it is easy to check if a system has a solution. Let

```
L=lead(C)
```

so that L will tell us if there is a leading 1 in the augmented column. Suppose there is a leading 1 in the augmented column. Look at the row which contains that leading 1; the first nonzero entry in that row occurs in the augmented column, and so the system is inconsistent. If there is no leading 1 in the augmented column, then the system is consistent.

MATLAB has an excellent general purpose solver for the system Ax=b. A solution is returned by x=A\b. If the system is consistent, then x=A\b will select a solution. If the system has more rows than columns, then x=A\b will select the "least squares solution." See *Least Squares* for what we mean by a "solution" to an inconsistent system. When we left our example the system Cx=d was inconsistent. Try

    x=C\d

What is C*x?

PROBLEMS

1. Find a solution to Ax=0 where A=magic(6). You may use rref. Assign random values to the free variables.

2. Construct A as follows: A=ones(6); A(:)=1:36. Now find a solution to Ax=0 using rref and assign random values to the free variables.

3. Write a MATLAB function B=elerow1(A,i,j) which switches rows i and j of A.

4. Write a MATLAB function B=elerow2(A,r,i) which multiplies row i of A by the scalar r.

5. Write a MATLAB function B=elerow3(A,r,i,j) which multiplies row i of A by the scalar r and adds it to row j.

6. Find the reduced row echelon form of magic(4). You should do this using the functions from problems 3,4 and 5. Check your work with rref.

7. Determine if Ax=b is consistent where A is the matrix created in problem 2 and

    (1) b=ones(6,1)

    (2) b=[0 0 0 0 0 1]'.

8. Write a MATLAB function c=con(A,b) which returns c=1 if the system Ax=b is consistent and c=0 if it is inconsistent.

9. Test con on the examples in problem 7.

10. We can write the solutions to $Ax = \vec{0}$ in the ***vector parametric form*** by finding vectors $v_1, \ldots, v_s$ such that every solution to $Ax = \vec{0}$ can be written as

$$x = t_1 v_1 + \cdots + t_s v_s$$

for some scalars $t_1, \ldots, t_s$. To find v_i let x(F)=e_i for $i = 1, \ldots, s$ and x(L)=-Ax. Here e_i is the ith standard basis vector (see page 9 of *Basics: Vectors and Matrices*) and s=sum(F). Try this on the matrix A in problem 2. You should find v_1, v_2, v_3, v_4.

11. Write a MATLAB function S=vp(A) which finds the vector parametric form of the solutions to Ax= 0 where S=$[v_1, \ldots, v_s]$. As a hint: the columns of the identity matrix are the standard basis vectors. What is A*S?

Building Matrices

ABSTRACT

In the *MATLAB Tutorial* you have seen the matrices generated by the functions `eye(n)`, `ones(n)`, `zeros(n)`, `rand(n)`, `hilb(n)`, and `magic(n)` as well as some matrix building functions. There are other matrix generating functions in MATLAB which will be used as needed. In this project you will see some matrices which will be useful for future projects. This project is also a good introduction to writing programs in MATLAB.

MATLAB COMMANDS

```
rref, diag, triu, tril, *, '
```

LINEAR ALGEBRA CONCEPTS

Elementary Matrices

BACKGROUND

A ***diagonal matrix*** is a square matrix of the form

$$\begin{bmatrix} a_{11} & 0 & \cdots & 0 \\ 0 & a_{22} & \ddots & \vdots \\ \vdots & \ddots & \ddots & 0 \\ 0 & \cdots & 0 & a_{nn} \end{bmatrix}$$

This can be easily created using the function `diag`. In MATLAB try

```
x=rand(5,1);
diag(x)
```

A ***superdiagonal matrix*** is a square matrix of the form

$$\begin{bmatrix} 0 & a_{1,2} & 0 & \cdots & 0 \\ 0 & 0 & a_{2,3} & \ddots & \vdots \\ \vdots & & \ddots & \ddots & 0 \\ \vdots & & & 0 & a_{n-1,n} \\ 0 & & \cdots & & 0 \end{bmatrix}.$$

Here again this can be created using `diag`. Try

```
diag(x,1)
```

A ***tridiagonal matrix*** has the form

$$\begin{bmatrix} a_{11} & a_{12} & 0 & \cdots & & 0 \\ a_{21} & a_{22} & a_{23} & & \ddots & \vdots \\ 0 & \ddots & \ddots & & \ddots & 0 \\ \vdots & \ddots & a_{n-1,n-2} & a_{n-1,n-1} & a_{n-1,n} \\ 0 & \cdots & 0 & & a_{n,n-1} & a_{n,n} \end{bmatrix}$$

31

In MATLAB try

```
A=rand(6)
triu(tril(A,1),-1)
```

A **Jordan Block** is a square matrix of the form

$$\begin{bmatrix} r & 1 & 0 & \ldots & 0 \\ 0 & r & 1 & \ddots & \vdots \\ \vdots & \ddots & \ddots & \ddots & 0 \\ \vdots & & \ddots & r & 1 \\ 0 & & \ldots & 0 & r \end{bmatrix}$$

where r is a scalar. A **companion matrix** is an $(n-1) \times (n-1)$ matrix of the form

$$\begin{bmatrix} -x_2/x_1 & -x_3/x_1 & \ldots & \ldots & -x_n/x_1 \\ 1 & 0 & \ldots & & 0 \\ 0 & 1 & \ddots & & \vdots \\ \vdots & & \ddots & 0 & 0 \\ 0 & & \ldots & 0 & 1 & 0 \end{bmatrix}$$

where $x = [x_1, x_2, \ldots, x_n]$ is any vector with $x_1 \neq 0$. The MATLAB function C=compan(x) finds the companion matrix. A **Vandermonde matrix** is a square matrix of the form

$$\begin{bmatrix} x_1^{n-1} & x_1^{n-2} & \ldots & x_1 & 1 \\ x_2^{n-1} & x_2^{n-2} & \ldots & x_2 & 1 \\ \vdots & \vdots & & \vdots & \\ x_n^{n-1} & x_n^{n-2} & \ldots & x_n & 1 \end{bmatrix}$$

where $x = (x_1, x_2, \ldots, x_n)$ is any vector. In MATLAB try

```
v=rand(4,1)
V=[v.^3, v.^2, v, ones(size(v))]
```

The MATLAB function V=vander(x) will find the Vandermonde matrix.

An inner product of two $n \times 1$ column vectors u and v is u·v=u'*v and this produces a 1×1 matrix. If we multiply u*v' we get an $n \times n$ matrix called the **outer product.** Try this in MATLAB

```
rand(5,1)*rand(1,5)
ones(5,1)*(1:5)
(1:5)'*ones(1,5)
```

Try rref on these matrices. What happens? The following matrix, pascal(n) is obtained from **Pascal's Triangle:**

$$P = \begin{bmatrix} 1 & 1 & 1 & 1 & 1 \\ 1 & 2 & 3 & 4 & 5 \\ 1 & 3 & 6 & 10 & 15 \\ 1 & 4 & 10 & 20 & 35 \\ 1 & 5 & 15 & 35 & 70 \end{bmatrix}$$

The rule for generating this is given by $P(i, j) = P(i, j - 1) + P(i - 1, j)$ for $i, j > 1$. The MATLAB function P=pascal(n) finds the Pascal matrix.

Let ele1(n, i, j) be the matrix resulting from switching the ith and j^{th} rows of the identity matrix I_n. Let ele2(n, r, i) be the result of multiplying the ith row of I_n by $r \neq 0$. Let ele3(n, r, i, j) be the result of multiplying the ith row of I_n by r and adding it to the jth row of I_n. These are the ***elementary matrices.*** Recalling the notation introduced for elementary row operations introduced in *Systems of Linear Equations* for the elementary row operations we have

$$\text{ele1}(n, i, j) = \text{elerow1}(I_n, i, j)$$
$$\text{ele2}(n, r, j) = \text{elerow2}(I_n, r, i)$$
$$\text{ele3}(n, r, i, j) = \text{elerow3}(I_n, r, i, j)$$

The next Theorem is a restatement of Theorem 4 in *Basics: Vectors and Matrices.*

Theorem. *Suppose A is an $m \times n$ matrix.*

    (1) *ele1$(m, i, j)A = $elerow1$(A, i, j)$,*
    (2) *ele2$(m, r, i)A = $elerow2$(A, r, i)$*
    (3) *ele3$(m, r, i, j)A = $elerow3$(A, r, i, j)$*

To illustrate this with an elementary row operation of type 3

```
A=magic(5)
E=eye(5)
E(2,:)=3*E(4,:)+E(2,:)
E*A
```

The ***Givens rotation matrix,*** $givrot(i, j, c, s)$, where $c^2 + s^2 = 1$ and $i \neq j$ (think of $c = \cos(\theta)$ and $s = \sin(\theta)$ for some angle θ) is found by starting with an identity matrix I and assigning $I(i, i) = c$, $I(i, j) = s$, $I(j, i) = -s$, and $I(j, j) = c$.

The ***Householder matrices*** have the economical description

$$H = I_n - (2/v^T v)vv^T$$

where v is a $n \times 1$ column vector. Notice that vv^T is an outer product. Try this

```
v=rand(6,1);
H=eye(6)-(2/(v'*v))*v*v';
```

We will call the following matrix list(n)

$$\begin{bmatrix} 1 & 2 & \cdots & n \\ n+1 & n+2 & \cdots & 2n \\ \vdots & & & \vdots \\ (n-1)n+1 & \cdots & \cdots & n^2 \end{bmatrix}$$

Try

```
A=zeros(5); A(:)=1:25,
A=A'
```

PROBLEMS

1. Make a 5×5 Jordan block with 3 on the diagonal.

2. Write a MATLAB function `J=jord(n,r)` which produces an $n \times n$ Jordan block matrix.

3. Letting `x=rand(1,5)` make a 4×4 companion matrix.

4. Let $c = \cos(\pi/6)$ and $s = \sin(\pi/6)$ and make all possible 3×3 Givens matrices.

5. Write a MATLAB function `G=givrot(n,i,j,c,s)` which produces an $n \times n$ Givens matrix. Using the matrices in problem 3 check your `givens`.

6. Write a MATLAB function `H=house(x)` which produces an $n \times n$ Householder matrix for the column vector `x`.

7. Make a Vandermonde matrix for the vector $(1, 2, 3, 4)$.

8. Make a matrix $A = \text{list}(6)$. Here is a start. Let `A=ones(6)` and assign `A(:)=1:36`.

9. Write a MATLAB function `L=list(n)` which produces the matrix list(n).

10. Write a MATLAB function `B=ele1(n,i,j)` which returns the $n \times n$ elementary matrix of type 1 which switches rows i and j of `eye(n)`. Let `E=ele1(5,2,4)` and check that `E=E'` and `E*E=eye(5)`.

11. Write a MATLAB function `B=ele2(n,r,i)` which returns the $n \times n$ elementary matrix of type 2 which multiplies row i of `eye(n)` by `r`. Let `E=ele2(5,2,5)` and check that `E=E'` and `E*ele2(5,1/2,5)=eye(5)`

12. Write a MATLAB function `B=ele3(n,r,i,j)` which returns the elementary matrix of type 3 obtained by multiplying row i of `eye(n)` by `r` and adding it to row j. Let `E=ele3(5,2,1,5)` and check that `E*ele3(5,-2,1,5)=eye(5)`.

Polynomials

ABSTRACT

A discussion on how to represent and manipulate polynomials in MATLAB.

MATLAB COMMANDS

```
conv, deconv, polyval, roots, +, *, \
```

LINEAR ALGEBRA CONCEPTS

Linear System of Equations

BACKGROUND

A polynomial is an expression of the form

$$f(x) = a_n x^n + a_{n-1} x^2 + \cdots + a_1 x + a_0.$$

Assuming $a_n \neq 0$, then we say that the **degree of** $f(x)$ **is** n and write this $\deg(f) = n$. Adding polynomials is similar to adding vectors. If $g(x) = b_m x^m + b_{m-1} x^{m-1} + \cdots + b_0$, then we add

$$f(x) + g(x) = (a_0 + b_0) + (a_1 + b_1)x + (a_2 + b_2)x^2 + \cdots$$

The polynomial $a_n x^n + a_{n-1} x^{n-1} + \cdots + a_1 x + a_0$ is represented in MATLAB by a vector $[a_n, a_{n-1}, \ldots, a_1, a_0]$. Notice the reverse role of the subscripts; a vector $a = [a_1, \ldots, a_n]$ represents the polynomial $a_1 x^{n-1} + a_2 x^{n-2} + \ldots + a_{n-1} x + a_n$. You need to be careful when adding polynomials of different degrees in MATLAB. Suppose that the vector a represents a polynomial $f(x)$ where $\deg(f) = n$ and b represents $g(x)$ where $\deg(g) = m$ and $m \leq n$. The vector which represents $f(x) + g(x)$ is a+[zeros(1,n-m),b] . Notice that

$$\deg(f(x) + g(x)) \leq \max(\deg(f(x), g(x))).$$

In MATLAB try

```
a=1:5, b=rand(1,3)
```

Their sum as polynomials is given by

```
a+[zeros(1,2),b]
```

Multiplying polynomials is slightly more complicated. Recall that if $f(x) = a_n x^n + \ldots + a_1 x + a_0$ and if $g(x) = b_m x^m + \ldots + b_1 x + b_0$, then $h(x) = g(x)f(x) = c_{m+n} x^{m+n} + \ldots + c_1 x + c_0$ where the coefficients are given by

$$c_k = \sum_{i=0}^{k} a_i b_{k-i}.$$

MATLAB will compute $f(x)g(x)$ with conv . Try

```
conv(a,b)
```

35

The formula for c_k is called a "convolution" hence the name conv for the MATLAB function. Notice that

$$\deg(f(x)g(x)) = \deg(f(x)) + \deg(g(x)).$$

Let P_n be the set of polynomials of degree $< n$. That is,

$$P_n = \{a_{n-1}x^{n-1} + \ldots + a_1 x + a_0 : a_0, \ldots, a_{n-1} \in R\}.$$

R^n is the set of vectors which represents P_n. The vectors $e_1, \ldots, e_n$ represents the polynomials $x^{n-1}, \ldots, x, 1$. For the purposes of addition and *scalar* multiplication, P_n is just like R^n.

If we are given a polynomial $f(x) = a_n x^n + \ldots + a_1 x + a_0$ and a scalar r we can find the value $f(r) = a_n r^n + \ldots + a_1 r + a_0$ in the usual way using the MATLAB functions $\hat{}, \backslash$ +, *. There is a classical method known as **Horner's method** for evaluating polynomials which uses the least number of +, *, \ operations and no $\hat{}$ operations. It is based on writing the polynomial as

$$f(r) = a_0 + r(a_1 + r(a_2 + \cdots + r(a_{n-1} + ra_n)\cdots).$$

The simple program to compute $s = f(r)$ is

```
s=0;
for i=n:-1:0
    s=s*r+a(i);
end
```

The MATLAB function polyval uses Horner's method, though the program looks slightly different since the MATLAB representation of the polynomial reverses the subscripts. Try polyval with

```
f=1:5
polyval(f,1)
sum(f)
```

Do you see why these are the same answer? Now try

```
polyval(f,0)
polyval(f,2)
```

A polynomial can be represented in **shifted power form** as

$$f(x) = a_0 + a_1(x - b) + a_2(x - b)^2 + \cdots + a_n(x - b)^n.$$

For any given b the coefficients $a_0, \ldots, a_n$ for the shifted power form can be computed from $f(r_0), \ldots, f(r_n)$ where the r_i's are any distinct values. This leads to a linear system of equations

$$f(r_0) = a_0 + a_1(r_0 - b) + a_2(r_0 - b)^2 + \cdots + a_n(r_0 - b)^n$$
$$f(r_1) = a_0 + a_1(r_1 - b) + a_2(r_1 - b)^2 + \cdots + a_n(r_1 - b)^n$$
$$\vdots$$
$$f(r_n) = a_0 + a_1(r_n - b) + a_2(r_n - b)^2 + \cdots + a_n(r_n - b)^n$$

The coefficients $a_0, a_1, \ldots, a_n$ are the unknowns which can be determined by solving this system. We use this form for the Taylor polynomials of a function in Calculus. A useful tool in working with polynomials is the **Division Algorithm for polynomials**

Theorem. *Division Algorithm. If $f(x)$ and $g(x)$ are polynomials and deg$(g) \geq 1$, then there exist polynomials $q(x)$ and $r(x)$ such that*

$$f(x) = g(x)q(x) + r(x) \qquad and \qquad deg(r) < deg(g).$$

There is a MATLAB function which finds the polynomials $q(x)$ and $r(x)$ from $f(x)$ and $g(x)$, namely `[q,r]=deconv(f,g)`. Try this

```
f=1:5, g=rand(1,4)
[q,r]=deconv(f,g)
```
Now check this with
```
conv(g,q)+r
```

The Division Algorithm is usually applied in the following way: Suppose that α is a scalar and $f(x)$ is a polynomial. Then by the Division Algorithm there are $q(x)$ and $r(x)$ such that

$$f(x) = (x - \alpha)q(x) + r(x) \text{ and } deg(r) = 0.$$

It follows that r is a constant. If α is a root, then when we substitute $x = \alpha$, we get $0 = f(\alpha) = (\alpha - \alpha)q(\alpha) + r = r$. Thus $f(x) = (x - \alpha)q(x)$. Now if $\alpha_1, \ldots, \alpha_n$ are all of the roots of $f(x)$ and $f(x)$ has x^n as the leading term, then $f(x) = (x - \alpha_1) \cdots (x - \alpha_n)$. In MATLAB we are going to evaluate the polynomial f created above at $\alpha = 1$. Remember that $[1, -1]$ represents the polynomial $x - 1$.
```
[q,r]=deconv(f,[1,-1])
```
Compare this with the result from `polyval`
```
polyval(f,1)
```

It is possible to evaluate a polynomial $f(x) = a_n x^n + a_{n-1} x^{n-1} + \cdots + a_1 x + a_0$ at an $n \times n$ matrix $x = A$. This is done as follows:

$$f(A) = a_n A^n + a_{n-1} A^{n-1} + \cdots + a_1 A + a_0 I_n.$$

The only possible surprise is the appearance of the I_n next to the constant term. Again, Horner's method can be applied to make this evaluation, and MATLAB's `polyvalm` (note the "m") will do this.

In some applications the vector x is the state of some system, and Ax is the state of the system after a unit of time. After m units of time $A^m x$ would describe the state of the system. See *Markov Chains* for more discussion of this type of model. Now we could evaluate A^m directly in MATLAB, but this could be time consuming even on a computer. If we can find a polynomial $f(x)$, where $f(A) = 0$, (this is the tricky part, see *Eigenvalues* for the Cayley-Hamilton Theorem, which addresses this question,) then we apply the Division Algorithm to get $q(x)$ and $r(x)$ such that

$$x^m = f(x)q(x) + r(x) \qquad \text{and} \qquad \deg(r) < \deg(f).$$

From this we get $A^m = f(A)q(A) + r(A) = r(A)$. Notice how $\deg(r)$ is always less than $\deg(f)$, regardless of the size of m. However, r can change if m is changed.

   We have only touched on the topic of **roots** or **zeros** here since one needs complex numbers to do justice to this topic. The reader should consult the projects *Complex Numbers* for more information. The MATLAB function `roots` will return approximations to the roots of a polynomial. This function uses a sophisticated method of finding the eigenvalues of a matrix to find the roots of a polynomial.

PROBLEMS

1. Use MATLAB to add and multiply the polynomials $x^3 + 2x - 1$ and $2x^5 + 3$. Now evaluate each of these polynomials at $x = -2$.

2. Find the shifted power form of $f(x) = x^3 + 2x - 1$ at $b = 1$. That is, find the coefficients for

$$f(x) = a_0 + a_1(x - 1) + a_2(x - 1)^2 + a_3(x - 1)^3.$$

You may choose $r_0 = 1$, $r_1 = 2$, $r_2 = 3$, $r_4 = 5$ or any distinct numbers.

3. Given the polynomial $h(x) = (x - 2)^3 + 3(x - 2)^2 - (x - 2) + 4$, how can you evaluate $h(2.1)$ using `polyval` directly on the shifted power form which is given here?

4. Use the MATLAB `roots` function to find the roots of the polynomials in problem 1.

5. Use MATLAB to show that $f(A) = 0$ where

$$A = \begin{bmatrix} 1 & 2 \\ 2 & 1 \end{bmatrix} \text{ and } f(x) = x^2 - 2x - 3$$

Now use `deconv` to compute A^{20}. Compare the answer with MATLAB's `A^ 20`. Using the fact that `f(A)=zeros(2)`, find a matrix B where `B*A=eye(2)`.

6. Write a MATLAB function `f=expan(r)` which finds the MATLAB representation for the polynomial

$$(x - r(1))(x - r(2)) \cdots (x - r(n))$$

where `r=[r(1),...,r(n)]`. Check expan by running `expan(r)` where `r=roots(f)` is the vector of roots found for each of the polynomials in problem 1.

7. Let `r=1:20` and let `f=expan(r)`. Find the roots of f (You know what they should be.) Now assign `f(2)=f(2)+`ϵ where $\epsilon = 10^{-k}$ for $k = 6, \ldots, 12$ and find the roots of each of these polynomials. The assignment `f(2)=f(2)+`ϵ changes the coefficient of x^{18} in the polynomial, it does not evaluate the polynomial. What happens? This is a famous example due to Wilkinson.

8. Let $f(x) = (x - 2)^{12}$. Here $f(x)$ is written in the shifted power form. Use `polyval` to compute the value at $x = 2.1$ without expanding the polynomial. This should be the expected answer. Now use expan to expand this polynomial out to show 13 terms and evaluate $f(2.1)$ again using polyval. This is one of the reasons for using the shifted power form.

9. To evaluate $f(t) = a_n t^n + a_{n-1} t^{n-1} + \ldots + a_0$ using Horner's method we get the following intermediate values for $s = f(t)$:

$$s_1 = a_n$$
$$s_2 = s_1 t + a_{n-1}$$
$$\vdots$$
$$s = s_{n+1} = s_n t + a_0$$

These linear equations lead to the following matrix equation:

$$
\begin{bmatrix}
1 & 0 & \cdots & \cdots & 0 \\
-t & 1 & 0 & & 0 \\
0 & -t & 1 & & \vdots \\
\vdots & & \ddots & \ddots & 0 \\
0 & \cdots & \cdots & -t & 1
\end{bmatrix}
\begin{bmatrix}
s_1 \\
s_2 \\
\vdots \\
s_{n+1}
\end{bmatrix}
=
\begin{bmatrix}
a_n \\
a_{n-1} \\
\vdots \\
a_0
\end{bmatrix}
$$

This equation can be set up and solved rapidly using MATLAB's A\b. Write a MATLAB function y=horner(a,t) which computes $f(t)$ in this manner. Check your horner against MATLAB's polyval on the polynomial f=1:10 evaluated at t=2.

10. The Binomial Theorem gives an explicit formula for the expansion of $(x + y)^n$. By setting $y = 1$ you can see the coefficients of this expansion using expan. Try this for $n = 3, 4, 5, 6$ and compare with the matrix pascal(6) (see page 33 in *Building Matrices* .)

11. Recall from Calculus the derivative of

$$f(x) = a_0 + a_1 x + a_2 x^2 + \ldots + a_n x^n$$

is

$$f'(x) = a_1 + 2a_2 x + 3a_3 x^2 + \ldots + n a_n x^{n-1}.$$

Write a MATLAB function d=der(f) which computes the representation of the derivative of the polynomial represented by f .

Graphics

ABSTRACT

This is a look at some of the graphics capabilities of MATLAB with an eye towards some of the geometry of linear algebra.

MATLAB COMMANDS

```
plot, mesh, axis, hold, clf, polyval,
meshgrid, print, *, \, +,-
```

LINEAR ALGEBRA CONCEPTS

Vector, Projection, Rotation, Reflection, Scale, Shear, Translation

BACKGROUND

This is primarily a graphics tutorial with some exercises at the end. MATLAB has two functions for graphics; for 2-dimensional plotting there is `plot`, and for 3-dimensional graphing, `mesh`. We consider `plot` first. Before beginning you may want to go through the demonstration program to view some of the graphics capabilities of MATLAB. To do that type `demo`.

Plotting Points

Set the axis with the command

```
axis([-6  6  -6  6])
```

I will explain more about `axis` later. You may want a grid imposed on your plots. If so, give the command `grid` after the plot command.

```
x=[3  5];
y=[1  2];
plot(x,y)
grid
```

The points $(3, 1)$ and $(5, 2)$ have been plotted and a line drawn between them. To suppress the line and just plot the points try

```
plot(x,y, '*')
```

```
plot(x,y, 'o')
```
Note: That is a lower case "oh."

```
x=[-2  1  0  -1]; y=[0  1  1  3];  plot(x,y)
```

The points $(-2, 0)$, $(1, 1)$, $(0, 1)$, $(-1, 3)$ have been plotted and connected by lines in this order.

Plotting Functions

The plot command for plotting a function $f(x)$ on an interval $[a, b]$ is the same;

```
plot(x,y)
```

where x and y are vectors and `y(i)=f(x(i))` for all `i`. Say we want a plot of $\sin(x)$. We first make a sampling vector x

```
x=-3:.1:3;
```

and then we find the function values

```
y=sin(x);
plot(x,y)
```

If we want to plot $y = 1/x$ with the same sampling vector, we have a problem since the sample x is a vector. We need a vector of ones to correct this.

```
y=ones(size(x))./x;
plot(x,y)
```

In general we choose a sampling vector

$$x = [a, a + res, a + 2 * res, \ldots, b]$$

which provides a plotting resolution, res. If we want to plot $n + 1$ equally spaced points on the interval $[a, b]$, then we choose $res = (b - a)/n$.

To plot $f(x) = x^2$, choose an interval, say $a = -1$, $b = 2$, and establish a sample vector:

```
a=-1; b=2; x=a:(b-a)/20:b;
y=x.^2;
plot(x,y)
```

You need to think of x as a vector when you are ready to create y. Now try $y = 1/x$.

```
y=ones(size(x))./x;
plot(x,y)
```

When you start plotting functions you will notice that MATLAB autoscales the axes. This means that MATLAB chooses the domain and range that will be plotted. Usually this is a big convenience. When we used the axis command earlier we were overiding the autoscaling feature. To release the axis that we have chosen, type

```
axis('auto')
```

Now try $y = 1/x$

```
plot(x,y)
```

That probably didn't work so well, on the other hand the auto-scaling works fine for $y = \sin(x)$

```
y=sin(x); plot(x,y)
```

You may prefer to take control of this, as I did when I directed you to type

```
axis([-6 6 -6 6]).
```

If v is a 1×4 vector, then the command axis(v) will be scale the x-axis to the interval [v(1),v(2)] and the y-axis to [v(3),v(4)]. To turn on the autoscaling again, just type

```
axis.
```

Plotting Polynomials

You can plot polynomials functions as above, but typing in the polynomial in this format is tedious. MATLAB represents a polynomial by a vector. The vector $a = [a_1, a_2, \ldots, a_n]$ represents the polynomial

$$a_1 x^{n-1} + a_2 x^{n-2} + \ldots + a_{n-1}x + a_n.$$

There are a variety of MATLAB functions which are intended to work on a vector as if it were a polynomial, *e.g.* `roots`, `conv`, `deconv`, and `polyval`. See *Polynomials* for more on these functions. For plotting we use `polyval` to get the polynomial values. Here is a polynomial plotting utility.

```
function polyplot(a,b,p,n)

% polyplot(a,b,p,n). This sets up the
% vector x=[a,a+res,...,a+n*res=b],
% where res=(b-a)/n, and y=f(x). The
% matrix p represents polynomials as
% the rows of p. All the polynomials
% are plotted. The default value of
% n is 50.

if nargin < 4, n=50; end
[k,l]=size(p);
x=(a:(b-a)/n:b)';
y=zeros(length(x),k);
for i=1:k,
    y(:,i)=polyval(p(i,:),x);
end
plot(x,y);
grid
```

The graphics data is usually stored separately from the other data in your session. If you want to impose another plot over an existing plot, type `hold` to hold the old screen while you give the new plot command. When you are ready to release the old plot, type `hold`. The command `figure` is used for managing the graphics windows.

Plotting Vectors

To plot the vector $(1, 2)$, that is, the vector with head at $(1, 2)$ and tail at $(0, 0)$ you type

```
plot([0; 1],[0; 2])
```

To plot the vector with tail at $(-1, 1)$ and head at $(2, 3)$ type

```
plot([-1;2],[1;3])
```

The following program is a vector plotting function which will plot vectors with their tails at a common point v. How does it work?

```
function vecplot(a,v)

% vecplot(a,v). The input a is a 2 by n
% matrix and the input v is a 2 by 1
% column vector. The columns of a are
% treated as the heads of 2 dimensional
% vectors with tails at v in the plot.
% If no v is specified, then the default
% value is the origin.

if nargin < 2, v=[0;0]; end
[m,n]=size(a);
[k,l]=size(v);
if m~=2,
    'ERROR:a has incorrect size',
    return,
end
if k~=2 | l~=1,
    'ERROR: v has incorrect size',
    return,
end
x=(v*ones(1,2*n))';
x(2:2:2*n,:)=a';
y=x(:,2);
x=x(:,1);
plot(x,y);
```

Here are the descriptions of some of the more common vector functions.

Rotation

The *rotation* through an angle θ can be accomplished by matrix multiplication.

$$R = \begin{bmatrix} \cos(\theta) & -\sin(\theta) \\ \sin(\theta) & \cos(\theta) \end{bmatrix}$$

For $v \in \mathbb{R}^2$, Rv is the rotated vector. In MATLAB

```
t=pi/3; R=[cos(t), -sin(t); sin(t), cos(t)];
```

```
v=[.5;1];
vecplot([v,R*v])
```

Projection

The ***projection of v onto w*** is given by

$$\text{proj}_w(v) = \frac{w \cdot v}{w \cdot w} w$$

```
w=[.5;.5]
vecplot([v,w]), hold
p=(w'*v)/(w'*w)*w
vecplot(p)
```

Did the plot change? Can you explain what happened?

The ***projection of v perpendicular to w*** is $v - \text{proj}_w(v)$.

```
q=v-p;
vecplot(q)
```

Reflection

The ***reflection of v across w*** is $-v + 2\text{proj}_w(v)$

```
u=-v+2*p;
vecplot(u)
```

The ***reflection of v perpendicular to w*** is $v - 2\text{proj}_w(v)$.

```
y=v-2*p;
vecplot(y)
```

Remove the "hold" by typing hold.

Scale

Scaling is multiplying a vector by a scalar.

```
w=2*v; vecplot(v), hold, plot(v,'*'),
vecplot(w)
```

Remove the "hold" by typing hold.

Shear

A ***shear*** is given by multiplying Sv where $S = \text{ele3}(2, r, i, j)$.

```
S=eye(2); S(1,:)=2*S(1,:);
u=S*v; vecplot([u,v])
```

Translation

Finally, there is the ***translation.*** If a vector is represented as a pair of points (v, w), with v as the tail and w as the head, then we translate by another vector u to the vector with head $v + u$ and tail $w + u$.

```
axis([0 2 0 2]),
vecplot(w,v), hold
u=[1;1]; vecplot(w+u,v+u)
```

Miscellaneous Plotting

Plot works well with curves given parametrically, $(x(t), y(t))$. Here you need only declare a domain for the parameter t, say t=-5:.1:5, let x=$x(t)$ and y=$y(t)$, and call plot(x,y).

It is also possible to plot in polar coordinates (θ, ρ). Suppose that $\rho = \rho(\theta)$ is a polar function. To plot the polar function $\rho = 1/\theta$ let

```
theta=pi/4:pi/20:4*pi;
rho=ones(size(theta)).\ theta;
polar(theta,rho).
```

MATLAB will allow plots with logarithmic scaling. Type help for information on loglog, semilogx and semilogy.

Three Dimensional Graphing

One of the primary functions is mesh which is applied to a matrix. The indices of the matrix are used as the xy plane and the matrix entries are graphed as as the z coordinate. Try

```
a=rand(10); mesh(a)
```

Applying a ***threshold*** to the matrix is a neat trick. Try

```
thresh=.2
b=(a<= thresh).*a, mesh(b)
```

For more fun try some other values for thresh. Enough fun. Say you want to graph a paraboloid like $z = x^2 + y^2$. The problem is to get the data into a matrix for graphing. MATLAB has a graphing utility meshgrid which facilitates this. If we let v be a vector of length n and w be a vector of length m, then

```
[x,y]=meshgrid(v,w)
```

creates two $m \times n$ matrices. The matrix x has v in every row and the matrix y has w(m:-1:1) in every column. The pairs (x(i,j),y(i,j)) give you the grid points in the grid formed by the vector v along the x-axis and the vector w along the y-axis. To plot the paraboloid first make a mesh domain

```
[x,y]=meshgrid(-5:.2:5,-5:.2:5)
```

```
z=x.^2 + y.^2; mesh(z)
```

Type `help mesh` for other pointers on changing the perspective of the graph. Some versions of MATLAB have a `rot90` command which will rotate the graph by 90 degrees.

Printing Graphics

To print the graphics screen type `print` to get a print of everything on the screen. Some systems support a command `print` to print the graphics screen. You can label your graphics with some text. See the commands `title`, `text`, `xlabel`, , and `ylabel`.

PROBLEMS

1. Try plotting the following functions
  (1) $\sin(x)$ on the interval $[-2\pi, 2\pi]$
  (2) x^3 on the interval $[-5, 5]$
  (3) $\frac{1}{1+x^2}$ on the interval $[-5, 5]$
  (4) $\frac{\sin(x)}{x}$ on the interval $[-3\pi, 3\pi]$
  (5) $\frac{\log(|x|)}{x}$ on the interval $[-5, 5]$

Try rescaling the axis to improve the view.

2. Try `polyplot` on the following polynomials. Try plotting them simultaneously.
  (1) x^2
  (2) $x^3 + x^2 + x + 1$
  (3) $x^5 - 5x^3 + 4x$

3. Let $x = (2, 0)$ and $y = (1, 1)$. Use `vecplot` to plot $x, y, x+y, x-y, 2x$.

4. Let x and y be the vectors from 3, and let S be the shear ele3(2, 2, 1, 2). Using `vecplot` plot x, y, Sx, Sy.

5. Let x and y be the vectors from 3, and let E be the scaling matrix ele2(2, 2, 1). Using `vecplot` plot x, y, Ex, Ey.

6. Let $v = (2, 0)$, $w = (1, 1)$, and p the projection of v onto w. Plot
```
   vecplot([v,w,p,v-p])
```
Now hold that screen and plot $v - p$ with its head at v and its tail at p.

7. Write a MATLAB program p=vecproj(w,v) which finds the projection of v onto w. Test it on the data in problem 6.

8. Let H = house(w) = $I_2 - (2/v^T v)vv^T$, then Hv is the reflection of v perpendicular to w and $-Hv$ is the reflection of v across w. For $v = (2, 0)$ and $w = (1, 2)$ plot, using `vecplot`, $v, w, Hv, -Hv$. See page 33 of *Building Matrices* for the definition of the Householder matrix.

9. The rotation of v by θ is r = givrot(n, i, j, c, -s) where $c = \cos(\theta)$ and $s = \sin(\theta)$. For v and w as in 7, plot $v, w, r * v, r * w$ using $i = 1, j = 2$ and $c = \cos(\pi/3)$ and $\sin(\pi/3)$. See page 'givens' of *Building Matrices* for the definition of of the Givens matrix.

10. Translate $u = (2, 0)$ and $v = (1, 1)$ by $w = (-3, 1)$. Plot all four vectors using `vecplot`.

11. Plot the following parametric curves.
  (1) $x(t) = t^3$ $y(t) = t^2$
  (2) $x(t) = t - \sin(t)$ $y(t) = 1 - \cos(t)$

12. Plot $x^2 + y^2 = 1$ by `t=0:pi/20:2*pi`, `x=cos(t)`, and `y=sin(t)`. Hold that plot. Now let `A=rand(2)` and transform the coordinates by

```
z=A*[x;y]
x1=z(1,:); y1=z(2,:);
plot(x1,y1)
```

13. Plot the following curves in polar coordinates

    (1) $\rho = 1 - 2\sin(\theta)$

    (2) $\rho = \cos(2\theta)$

14. Using `mesh` graph the following.

    (1) $z = \sin(\sqrt{x^2 + y^2})$

    (2) $z = y^2 - x^2$

    (3) $z - x^2 - y^2 = 1$

    (4) $z = x^3 - 3xy^2$

Bézier Curves

ABSTRACT

A Bézier curve is a parameterized polynomial curve for approximating a sequence of data points. Bézier curves have been used in computer aided design and manufacturing. This project makes extensive use of MATLAB's polynomial representation which is covered in *Polynomials*.

MATLAB COMMANDS

    polyval, plot, hold, *

LINEAR ALGEBRA CONCEPTS

Linear Combination

BACKGROUND

Bézier curves are used by manufacturers since they offer an elegant way to specify instructions for milling and contouring machines. Designers appreciate the economy of being able to get an attractive curve by designating a few **control points** to guide the curve. Given points $(x_0, y_0), (x_1, y_1), \ldots, (x_n, y_n)$ a **Bézier curve** is a parameterized polynomial curve, $(x(t), y(t))$, given by

$$x(t) = \sum_{i=0}^{n} x_i B_i^n(t)$$

$$y(t) = \sum_{i=0}^{n} y_i B_i^n(t)$$

where the $B_i^n(t)$ are Bernstein polynomials. We need a little background on Bernstein polynomials.

Bernstein Polynomials

The **Bernstein polynomials of degree n** are defined by

$$B_i^n(x) = C(n, i)(1 - x)^{n-i} x^i$$

where $i = 0, \ldots, n$ and $C(n, i)$ is the binomial coefficient. The binomial coefficients can be defined directly by $C(n, i) = \frac{n!}{i!(n-i)!}$ or we can appeal to the matrix $P = \text{pascal}(n)$ from page 32 of *Building Matrices*, since $C(n, i) = P(n - i + 1, i + 1)$. This definition is not quite as off-the-wall as it appears. If p is the probability of an event, then the binomial distribution gives the probability of the event occurring exactly i times in n independent trials as $B_i^n(p)$.

The Bernstein polynomials of degree 1:

$$B_0^1(t) = -t + 1$$
$$B_1^1(t) = t$$

The Bernstein polynomials of degree 2:

$$B_0^2 = (1-t)^2 = t^2 - 2t + 1$$
$$B_1^2 = -t^2 + t$$
$$B_2^2 = t^2$$

The Bernstein polynomials of degree 3.

$$B_0^3(t) = (1-t)^3 = -t^3 + 3t^2 - 3t + 1$$
$$B_1^3(t) = 3(1-t)^2 t = 3t^3 - 6t^2 + 3t$$
$$B_2^3(t) = 3(1-t)t^2 = -3t^3 + 3t^2$$
$$B_3^3(t) = t^3$$

One of the items we will want is a recurrence relation for $B_i^n(x)$. We first need one for $C(n, i)$. Recall from the definition of $P = \text{pascal}(n)$; $P(i, j) = P(i, j-1) + P(i-1, j)$. Translating this to C we get $C(n, i) = C(n-1, i-1) + C(n-1, i)$. This yields the recurrence relation for Bernstein polynomials.

Theorem. *For all n, $B_0^n(t) = (1-t)^n$ and $B_n^n(t) = t^n$. For $n > 1$ and $0 < i < n$*

$$B_i^n(t) = (1-t)B_i^{n-1}(t) + tB_{i-1}^{n-1}(t).$$

PROOF:

$$\begin{aligned} B_i^n(t) &= C(n, i)(1-t)^{n-i} t^i \\ &= (C(n-1, i) + C(n-1, i-1))(1-t)^{n-i} t^i \\ &= (1-t)C(n-1, i)(1-t)^{n-i-1} t^i + \\ &\quad tC(n-1, i-1)(1-t)^{n-i} t^{i-1} \\ &= (1-t)B_i^{n-1}(t) + tB_{i-1}^{n-1}(t). \end{aligned}$$

∎

The program bern.m lists the Bernstein polynomials of degree n as the columns of an $(n+1) \times (n+1)$ matrix. The program is based on the recurrence relation given above.

```
function b=bern(n)

% b=bern(n). Each row of b is an n-th degree
% Bernstein polynomial.

b=[-1 1;1 0];
if n > 1,
  for k=3:n+1;
    a=zeros(k);a(1,k)=1;
    a(:,1)=[0;b(:,1)]-[b(:,1);0];
    for j=2:k-1,
      a(:,j)=[0;b(:,j)]-[b(:,j);0];
      a(:,j)=a(:,j)+[b(:,j-1);0];
    end;
    b=a;
  end
end
```

Type in bern.m and try a few values in MATLAB

```
bern(3)
bern(5)
```

Bézier Curves

Suppose that we are given $(x_0, y_0), (x_1, y_1), \ldots, (x_n, y_n)$ and we wish to *approximate* these points with a Bézier curve. The Bézier curve is a parameterized polynomial curve, $(x(t), y(t))$, given by

$$x(t) = \sum_{i=0}^{n} x_i B_i^n(t)$$

$$y(t) = \sum_{i=0}^{n} y_i B_i^n(t)$$

where $0 \le t \le 1$. If we let $t = 0$ we get $x(0) = x_0$ and $y(0) = y_0$. Similarly, if we let $t = 1$ we get $x(1) = x_n$ and $y(1) = y_n$. Other than that the curve does not necessarily pass through the points.

To convert this definition into a matrix problem, consider the polynomials $B_i^n(t)$ as being column vectors as they are in the matrix B=bern(n) and let x be the column vector with entries x_i, and y the column vector with entries y_i. Then $x(t)$ is given by

$$\sum_{i=0}^{n} \text{x(i)B(:,i)} = \text{B*x}$$

and $y(t)$ is given by

$$\sum_{i=0}^{n} \text{y(i)B(:,i)= B*y}$$

The program `bezier.m` utilizes this to determine the Bézier curve.

```
function [xb,yb]=bezier(x,y,res)

% [xb,yb]=bezier(x,y,res). Computes coordinates for
% a Bezier curve approximating the data given by
% the points (x,y). "res" determines the number of
% plotting points. The default value of "res" is 100.

if nargin < 3, res=100; end
x=x(:); y=y(:);
[m,n]=size(x);
poly=bern(m-1);
xpoly=poly*x; ypoly=poly*y;
t=0:1/res:1;
xb=polyval(xpoly,t); yb=polyval(ypoly,t);
```

The program bezier will return vectors xb and yb ready for plotting. For the rest of the project we will try this function on sample data. Begin by looking at two points. $P_0 = (0, 0)$ and $P_1 = (1, 1)$. We know that the curve will pass through the first and last points and that the degree of the polynomials will be 1. So we should get a line.

```
x=[0;1]; y=[0;1];
[xb,yb]=bezier(x,y);
plot(xb,yb)
```

No real surprise here. Now try three points on a line.

```
x=[0 1 2]'; y=[0 1 2]';
[xb,yb]=bezier(x,y);
plot(xb,yb)
```

What about three noncollinear points?

```
x=[0 1 2]'; y=[0 1 0]';
[xb,yb]=bezier(x,y);
plot(x,y), hold
plot(xb,yb),
```

Let's try a jagged edge. Be sure to remove the "hold" by typing hold.

```
x=[0 1 2 3 4]'; y=[0 1 0 2 1]';
plot(x,y), hold
```

```
[xb,yb]=bezier(x,y);
plot(xb,yb)
```

Now let's try something more daring. Be sure to remove the "hold."

```
x=[0 1 3 2 1 3 4]'; y=[0 0 -1 -2 -1 0 0]';
plot(x,y), hold
[xb,yb]=bezier(x,y);
plot(xb,yb)
```

Now open up the inside of the curve a bit. Remove "hold."

```
x(3)=4;x(5)=0;
plot(x,y), hold
[xb,yb]=bezier(x,y);
plot(xb,yb)
```

We can make a French curve by specifying just a few points.

```
x=[0 -2 5 5 -5 0]'; y=[0 -3 5 -3 -3 0]';
plot(x,y), hold
[xb,yb]=bezier(x,y);
plot(xb,yb)
```

A serious problem with the Bézier curves is that the degree of the curve goes up with the number of data points. You may have noticed that bezier took longer with more control points.

PROBLEMS

1. Find the Bézier curve and plot the curve with the input data

```
x=[-2 -.5 -.5 0 .5 1]
y=[0 -1 1 0 2 0].
```

Now switch (x(2),y(2)) and (x(3),y(3)) and plot it again. You can see that the order that the points are placed makes a big difference.

2. Find the Bézier curve and plot the curve with the input data

```
x=[0 1 2 3 4 5 4 3 4 5 6]
y=[-1 2 1 3 2 -1 -2 -1 1 1 3].
```

3. One of the useful features of a parameterized curve is that you can manipulate it like a vector. For example you can apply a rotation easily, see page 43 of *Graphics* for background on rotations. Let xb and yb be the graphing vectors for the curve in problem 2.

```
t=pi/6;
r=[cos(t),-sin(t);sin(t),cos(t)];
z=r*[xb;yb];
xr=z(1,:); yr=z(2,:); plot(xr,yr)
```

4. Suppose we place the control points on a circle, will they form a circle?

```
t=[0:4,0]*2*pi/5';  x=cos(t);y=sin(t);
plot(x,y), hold
[xb,yb]=bezier(x,y);  plot(xb,yb)
```

Move the starting point inside the pentagon. Remove the "hold."

```
x=[0;x;0];  y=[0;y;0];
plot(x,y)
[xb,yb]=bezier(x,y);  plot(xb,yb)
```

5. A reasonable approximation to an arc of a circle can be made with four control points and a cubic Bézier curve. First note what happens when we choose

```
x=[1 1 0]';  y=[0 1 1]';
plot(x,y), hold
[xb,yb]=bezier(x,y);  plot(xb,yb)
```

To see how well you have done plot a circle.

```
t=0:pi/20:pi/2;
xcirc=cos(t);  ycirc=sin(t);  plot(xcirc,ycirc)
```

We can do better by adding some control points. Suppose that we add points $(s, 0)$ and $(0, s)$ with s to be determined. We would like the curve to have values $x(.5) = \cos(\pi/4)$ and $y(.5) = \sin(\pi/4)$. Thus from the definition of a Bézier curve

$$x(t) = x_0 B_0^3(t) + x_1 B_1^3(t) + x_2 B_2^3(t) + x_3 B_3^3(t)$$

and since x=[1 1 s 0] and y=[0 s 1 1] this gives

$$x(.5) = \cos(\pi/4) = B_0^3(.5) + B_1^3(.5) + s B_2^3(.5)$$
$$= 4(.5)^3 + 3s(.5)^3$$

```
s=(cos(pi/4)-4*.5^3)/(3*.5^3);
x=[1 1 s 0];  y=[0 s 1 1];
[xb,yb]=bezier(x,y);
```

Remove the "hold" and plot.

```
plot(x,y), hold, plot(xb,yb),
plot(xb,yb)
```

Finish the rest of the circle by rotating as in problem 3.

6. Using the technique of problem 5 approximate the parabola $y = x^2$ on the interval $[0, 1]$. We need to find points $(r, 0)$ and $(1, s)$ so that for $x = [0, r, 1, 1]$ and $y = [0, 0, s, 1]$

$$x(t) = x_0 B_0^3(t) + x_1 B_1^3(t) + x_2 B_2^3(t) + x_3 B_3^3(t)$$

and

$$y(t) = y_0 B_0^3(t) + y_1 B_1^3(t) + y_2 B_2^3(t) + y_3 B_3^3(t)$$

To determine r and s it is easiest to try to figure out, as we did in problem 5 what values we want for $x(.5)$ and $y(.5)$. In this case using the point of intersection between $y = x^2$ and $y = -x + 1$ (draw a picture to see why this is reasonable) works well. You can use the MATLAB function roots to get the point of intersection.

7. Using the technique of problem 5 approximate the ellipse $x^2 + y^2/4 = 1$ in the first quadrant.

8. Using the technique of problem 5 approximate the cusp $y = x^{2/3}$ on the interval $[-1, 1]$.

9. Plot the Bernstein polynomials of degree 3 on the interval $[0, 1]$. Now do the Bernstein polynomials of degree 4. This can be easily done with polyplot from page 42 of *Graphics*.

10. Find a data set for which the Bézier curve is the profile of a sleek automobile of your own design.

Lagrange Interpolation

ABSTRACT

Given any points in $\mathbb{R}^2$ there is a polynomial function which passes through these points. We will show how to find this polynomial.

MATLAB COMMANDS

`\`, `plot`

LINEAR ALGEBRA CONCEPTS

Linear System of Equations

BACKGROUND

We consider the **interpolation problem.** Given data points $(a_1, b_1), \ldots, (a_n, b_n)$ where the a_i's are distinct. The x-coordinates $a_1, \ldots, a_n$ are called **knot points** or simply **knots.** Our goal is to find the polynomial, $f(x)$ of degree $n-1$, such that $f(a_i) = b_i$ for $i = 1, \ldots, n$. This is easy and can be done by solving a matrix equation.

Let $f(x) = c_{n-1}x^{n-1} + \ldots + c_1 x + c_0$ where the c_i's are unknowns. The **Lagrange interpolating polynomial** is found by solving for the coefficients. Then we get for $i = 1, \ldots, n$

$$f(a_i) = c_{n-1}a_i^{n-1} + \ldots + c_1 a_i + c_0 = b_i$$

and so

$$\begin{bmatrix} a_1^{n-1} & \cdots & a_1 & 1 \\ a_2^{n-1} & \cdots & a_2 & 1 \\ \vdots & & \vdots & \\ a_n^{n-1} & \cdots & a_n & 1 \end{bmatrix} \begin{bmatrix} c_{n-1} \\ \vdots \\ c_1 \\ c_0 \end{bmatrix} = \begin{bmatrix} b_1 \\ b_2 \\ \vdots \\ b_n \end{bmatrix}$$

The coefficient matrix is the Vandermonde matrix on page 32 from *Building Matrices*. When the a_i are distinct, the Vandermonde matrix is invertible and so the system $Ax = b$ can be solved. Let's try this.

    a=(1:4)'; b=[0 1 0 2]';

This gives us four points.

    plot(a,b,'*'), hold

Form the Vandermonde matrix.

    A=[a.^3, a.^2, a, ones(a)],

    c=A\b

To plot we use `polyplot` from page 42 of *Graphics*

    polyplot(1,4,c')

A different form of the Lagrange polynomial, the **Newton form,** is written

$$f(x) = c_0 + c_1(x - a_1) + c_2(x - a_1)(x - a_2) + \cdots +$$
$$c_{n-1}(x - a_1) \cdots (x - a_{n-1})$$

where $c_0, \ldots, c_{n-1}$ are to be determined. The coefficients of the Newton form are not the same as the coefficients of the standard form of the Lagrange polynomial. This can be written more compactly as

$$f(x) = \sum_{k=0}^{n-1} c_k \prod_{i=1}^{k} (x - a_i)$$

We can use a matrix equation to determine the coefficients of the Newton form by substituting in the data points:

$$b_1 = c_0$$
$$b_2 = c_0 + c_1(a_2 - a_1)$$
$$\vdots$$
$$b_n = \sum_{k=0}^{n-1} c_k \prod_{i=1}^{k} (a_n - a_i)$$

This gives the matrix equation

$$\begin{bmatrix} 1 & 0 & \cdots & & 0 \\ 1 & a_2 - a_1 & & & \vdots \\ \vdots & & \ddots & & 0 \\ 1 & a_n - a_1 & \cdots & & \prod_{i \leq n-1}(a_n - a_i) \end{bmatrix} \begin{bmatrix} c_0 \\ c_1 \\ \vdots \\ c_{n-1} \end{bmatrix} = \begin{bmatrix} b_1 \\ b_2 \\ \vdots \\ b_n \end{bmatrix}$$

where the (k, l) entry for $k \geq l$ is $\prod_{i<k}(a_l - a_i)$. Since this is a lower triangular matrix we can find the solution using forward substitution. If you have `forsub` (see page 92 of *Systems of Linear Equations*) this is quite easy. The Newton form is a different representation than the MATLAB representation of a polynomial and consequently the MATLAB polynomial functions like `polyval` cannot be used on the Newton form.

Another approach to the Lagrange polynomial is through the ***Lagrange basis polynomials***. Given distinct knots $a_1, \ldots, a_n$ form the Lagrange basis polynomials

$$L_1(x) = \prod_{j \neq 1} \frac{(x - a_j)}{(a_1 - a_j)}$$

Notice that $L_1(a_j) = 0$ if $j \neq 1$ and $L_1(a_j) = 1$ if $j = 1$. Now more generally, form

$$L_i(x) = \prod_{j \neq i} \frac{(x - a_j)}{(a_i - a_j)}$$

And again observe that $L_i(a_j) = 0$ if $i \neq j$ and 1 if $i = j$. To form the Lagrange interpolating polynomial for the data $(a_1, b_1), \ldots, (a_n, b_n)$ take the polynomials $L_1, \ldots, L_n$ and define

$$f(x) = \sum_{i=1}^{n} b_i L_i(x).$$

Notice that $f(a_j) = \sum_{i=1}^{n} b_i L_i(a_j) = b_j L_j(a_j) = b_j$.

Approximating a Function

We find an interpolating polynomial for a function $f(x)$ by selecting knots $x_1, \ldots, x_n$ and letting $y_1 = f(x_1), \ldots, y_n = f(x_n)$. The interpolating polynomial through this data approximates the function $f(x)$. There is some freedom in choosing the knots $x_1, \ldots, x_n$. One way is **regularly spaced** data on the interval $[a, b]$ given by

$$x_i = a + (\frac{b-a}{n})i$$

for $i = 0, \ldots, n$. Another choice of knots is given by the **Chebyshev points**

$$x_i = \frac{a+b}{2} + \frac{a-b}{2} \cos(\frac{(2i-1)\pi}{2n})$$

for $i = 1, \ldots, n$.

PROBLEMS

1. Let a=1:5 and b=[-1,2,0,1,4]. Set up the Vandermonde matrix and use it to find the coefficients of the Lagrange interpolating polynomial. Graph it to see how it passes through the data.

2. Using the data from problem 1, and the formula

$$f(x) = \sum_{i=1}^{n} b_i L_i(x)$$

find the Lagrange interpolating polynomial. Does it agree with the polynomial in problem 1?

3. Using the data in problem 2, find the Newton form of the interpolating polynomial. You may use forsub see page 92 of *Systems of Linear Equations* or you can use \ to solve for the coefficients.

4. Write a MATLAB function c=lagrange(a,b) which returns the coefficents of the Lagrange interpolating polynomial through the data (a(1),b(1)), ..., (a(n),b(n)) where n=length(a) =length(b). Check your program on the data in problem 1.

5. Find the interpolating polynomial for $\sin(x)$ using 6 regularly spaced knot points on the interval $[0, \pi]$. Plot both $\sin(x)$ and the interpolating polynomial.

6. Find the interpolating polynomial for $\frac{1}{25x^2+1}$ using 6 regularly spaced knot points on the interval $[-1, 1]$. Plot both $\frac{1}{25x^2+1}$ and the interpolating polynomial. This is an example due to Runge. See if you can improve the situation using more knots, say 10 or 12.

7. Find the interpolating polynomial for $\frac{1}{25x^2+1}$ using 6 Chebyshev knots on the interval $[-1, 1]$. Plot both $\frac{1}{25x^2+1}$ and the interpolating polynomial. Try increasing the number of knots to 10.

8. Find an interpolating polynomial for abs(x) on $[-1, 1]$ using 6 regularly spaced knots. Try it with 6 Chebyshev knots. Plot the interpolating polynomial over abs(x) in each case.

9. Do problem 8 for the pulse function

$$f(x) = \begin{cases} 0 & \text{if } -1 \le x < 0 \\ 1 & \text{if } 0 \le x \le 1 \end{cases}$$

How does it change if we increase the number of data points to 10?

10. Let a=1:5 and form the Lagrange polynomials $L_1(x), \ldots, L_5(x)$ using a_i=a(i). You might want to use the function expan on page 38 from *Polynomials* to get the standard representation of these polynomials. Now graph these polynomials on the interval $[0, 6]$. Notice how $L_i(x)$ is 1 at $x = a(i)$ and 0 for $x = a(j)$ for $j \ne i$.

11. MATLAB has a feature flops which counts the number of arithmetic operations used. We know that many algorithms in linear algebra have flop counts given as polynomial expressions of the size of the inputs. For example matrix multiplication of $n \times n$ matrices uses $c_1 n^3 + c_2 n^2 + c_3 n + c_4$ flops for some coefficients c_i which can be determined. In fact we can determine this easily by generating flop counts for matrix multiplications A*B where A and B are random $n \times n$ matrices for $n = 3, 5, 8, 10$.

```
t=[3 5 8 10];
for i=t,a=rand(i);c=rand(i);flops(0),
a*c;b=[b,flops];end
```

Apply Lagrange interpolation to the resulting data. Try this. How well does the interpolating polynomial predict for $n = 15$? $n = 20$?

COMMENTS

The MATLAB function polyfit does least squares fitting. By taking the degree in polyfit to be $n - 1$ when length$(a) = n$ you will get the interpolating polynomial.

Complex Numbers

ABSTRACT

A discussion on complex numbers and how MATLAB handles them.

MATLAB COMMANDS

```
plot, mesh, meshgrid, +, *, ', conj, norm, abs,
roots, real, imag, exp, sign
```

LINEAR ALGEBRA CONCEPTS

Transpose

BACKGROUND

A *complex number* is a number of the form $z = a + bi$ where $i = \sqrt{-1}$ and a and b are real numbers. To see a complex number in MATLAB, simply type

```
z=2+3*i
```

i is a built-in variable in MATLAB like pi. The real number a is called the *real part* of z and the real number b is called the *imaginary part.* If you want to access a and b in MATLAB just type

```
real(z)
```

```
imag(z)
```

Arithmetic with complex numbers is quite simple

$$(a + bi) + (c + di) = (a + c) + (b + d)i$$

and

$$(a + bi) \cdot (c + di) = ac + adi + bci + bdi^2 = (ac - bd) + (ad + bc)i$$

There is a new operation with complex numbers, *conjugation.* For $z = a+bi$, the *conjugate,* $\overline{z} = a - bi$. In MATLAB, the conjugate is handled by either ' or conj

```
(2+3*i)'
```

```
conj(2+3*i)
```

The function `A'` performs the matrix transpose on matrices with real number entries. Actually, `A'` simultaneously performs the transpose of A and conjugates any complex entries in A making it the "conjugate transpose." The function `A.'` is the transpose without the conjugate. The function `conj` will conjugate the elements of a matrix without transposing. When we multiply

$$(a + bi)(\overline{a + bi}) = (a + bi)(a - bi) = a^2 + b^2$$

we get a real number. Using this we get a simple formula for division of complex numbers.

$$\frac{a + bi}{c + di} = \frac{(a + bi)\overline{(c - di)}}{(c + di)\overline{(c - di)}} = \frac{1}{c^2 + d^2}((ac + bd) + i(ad + bc))$$

Conjugation has many pleasant properties. Here is a list some of them here without proof.

$$\overline{\overline{z}} = z$$

$$\overline{z_1 \pm z_2} = \overline{z_1} \pm \overline{z_2}$$

$$\overline{z_1 z_2} = \overline{z_1}\, \overline{z_2}$$

$$\overline{z_1/z_2} = \overline{z_1}/\overline{z_2}$$

You may ask if we really need complex numbers. The solutions to the polynomial equation $x^2 = -1$ are the complex numbers $\pm i$. More generally, the solutions to the quadratic equation $ax^2 + bx + c = 0$ are given by the quadratic formula

$$\frac{-b \pm \sqrt{b^2 - 4ac}}{2a}$$

which produces complex numbers when $b^2 - 4ac < 0$. If you have worked through *Polynomials,* I suggest that you try an experiment like

```
f=1:10
```

```
roots(f)
```

This gives you some sense of how easily complex numbers appear as roots of polynomials. If you do this experiment you will notice that if z occurs as the root, then so does $\overline{z}$. This is always true if the polynomial has *real* coefficients.

Graphing and Polar Form

A complex number $z = a + bi$ can be graphed as the pair (a, b) in the plane. When we do this we refer to the plane as the ***complex plane;*** the x−axis is called the ***real axis,*** and the y−axis is called the ***imaginary axis.*** MATLAB will graph in the complex plane automatically.

```
z=2+3*i
```

```
plot(z,'*')
```

The distance from 0 to $z = a + bi$ is given by the distance formula

$$|z| = \sqrt{a^2 + b^2}$$

and is referred to as the ***norm of*** z. This absolute value notation does not collide with our usual use of this notation, for if z is a real number, then $b = 0$ and so $|z| = \sqrt{a^2} = |a|$. Notice that in terms of the conjugate, $z\overline{z} = |z|^2$. In MATLAB either the function abs(z) or norm(z) will compute the norm of a complex number.

You may recall polar coordinates from Calculus. If θ is the angle between the positive x−axis and the segment connecting $(0, 0)$ and (a, b), and r is the distance from $(0, 0)$ to (a, b), then by a little trigonometry $a = r\cos(\theta)$ and $b = r\sin(\theta)$. Substituting these expressions for a and b into $z = a + bi$ we get the ***polar form***

$$z = r(\cos(\theta) + i\sin(\theta))$$

If you want to find the polar form, just note that $r =$ norm(z) and $(1/r)z = \cos(\theta) + i \sin(\theta)$. MATLAB's sign function is defined for complex numbers as sign$(z) = \cos(\theta) + i \sin(\theta)$, thus $z =$ norm(z) sign(z). Now you can find θ using real, imag, and acos and asin.

Using the trig identities $\sin(\theta + \phi) = \sin(\theta)\cos(\phi) + \cos(\theta)\sin(\phi)$ and $\cos(\theta + \phi) = \cos(\theta)\cos(\phi) - \sin(\theta)\sin(\phi)$ we get

Theorem. *DeMoivre's Theorem. If $z = r(\cos(\theta) + i \sin(\theta))$ and $w = s(\cos(\phi) + i \sin(\phi))$ are complex numbers written in polar form, then*

$$zw = rs(\cos(\theta + \phi) + i \sin(\theta + \phi)).$$

It follows that for any integer $n \geq 0$

$$z^n = r^n(\cos(n\theta) + i \sin(n\theta))$$

Try the following little experiment. Let

```
z=2+3*i;
z=sign(z)
```

Thus norm$(z)=1$. Now let

```
w=[z,z^2,z^3,z^4,z^5,z^6, z^7];
plot(w)
```

Each of the vertices of the resulting figure is one of the powers of z from above. Notice how the angle between adjacent vertices is the same. This is exactly as predicted by DeMoivre's Theorem.

Consider the equation $x^n = 1$. The roots of this equation are called the nth *roots of unity.* We know one root, $x = 1 = \cos(0) + i \sin(0)$. The others are found by dividing up the unit circle into n arcs. For $k = 0, \ldots, n - 1$ define

$$z_k = \cos(2k\pi/n) + i \sin(2k\pi/n)$$

and notice that by DeMoivre's Theorem

$$z_k^n = \cos(2nk\pi/n) + i \sin(2nk\pi/n) = \cos(2\pi) + i \sin(2\pi) = 1.$$

Thus for each k the z_k defined above is a root of $x^n = 1$ and so these constitute the n roots of unity. You can see the nth roots of unity in the complex plane.

```
z=cos(2*pi/5)+i*sin(2*pi/5)
w=[z,z^2,z^3,z^4,z^5, z^6]
plot(w)
```

Notice that if $z = \cos\theta + i\sin\theta$, then $|z| = \sqrt{\cos^2\theta + \sin^2\theta} = 1$, so that for $z \neq 0$, $|z|$ always lies on the unit circle.

Exponential Form

Suppose that $z = a + bi$. The extension of the exponential e^x to the complex numbers is defined by

$$e^z = e^a e^{bi}$$

which is natural enough, but we also define e^{bi} by

$$e^{bi} = \cos(b) + i\sin(b)$$

This definition is justified by the power series for e^x, $\sin(x)$, and $\cos(x)$. MATLAB's exp finds the exponential of a complex number automatically. Try

```
exp(2+3*i)
```

Every complex number z can be written in polar form

$$z = r(\cos(\theta) + i\ \sin(\theta))$$

where $r > 0$ and we get the definition of the *exponential form* $z = re^{i\theta}$. Since $r > 0$ we can solve $r = e^u$ for u and substituting this in $x = e^u e^{i\theta} = e^{u+i\theta}$. Thus every complex number can be written in the form $z = e^w$ for some complex w.

The amazing Euler Formulas come from the definition of the complex exponential:

$$e^{ix} = \cos(x) + i\sin(x)$$
$$e^{-ix} = \cos(x) - i\ \sin(x)$$

and letting $x = \pi$ we get

$$e^{\pi i} = \cos(\pi) + i\ \sin(\pi) = -1$$

Try it in MATLAB!

```
exp(i*pi)
```

Now look at $e^{i\theta} = \cos(\theta) + i\sin(\theta)$ where $0 \le \theta \le 2\pi$. In MATLAB type

```
x=0:pi/50:2*pi;
y=exp(i*x);
plot(x,y)
```

Surprised? It comes directly from the definition.

In exponential form the roots of unity are easily described; for each $k = 0, \dots, n-1$ let $z_k = e^{2k\pi/n}$. Now

$$z_k^n = e^{2kn\pi/n} = e^{2k\pi} = 1.$$

Graphing Complex Functions

Many functions on the real numbers can be extended to the complex numbers just as we did with the exponential function. This is most easily accomplished when we have a power series expansion for the real function. The built-in functions in MATLAB can all take complex numbers as arguments. To see what these functions look like you can get a graph using the mesh feature. Suppose that you want a graph of e^z on the rectangle $z = x + yi$ where $-2 \leq x \leq 2$ and $-2 \leq y \leq 2$. Type

```
[x,y]=meshgrid(-2:.2:2,-2:.2:2);
z=x+i*y
r=exp(z)
mesh(r)
```

With a complex function the output will also be a complex number. You have already seen this when you tried `exp(2+3*i)`. In the previous mesh graph only the real part of the output was displayed. To see the complex part

```
mesh(imag(r))
```

These functions are remarkably beautiful.

PROBLEMS

1. Let `a=magic(4)` and `b=rand(4)`. Perform the following operations:

```
c=a+i*b
real(c)
imag(c)
c'
conj(c)
(2+3*i)*c
abs(c)
```

Check the claim that `z=norm(z)*sign(z)` on `z=2+3*i`.

2. Let $z = 1 - 2i$. Solve for r and θ to get the polar form $z = r(\cos(\theta) + i\sin(\theta))$.

3. Let $z = 1 + i$. Solve for r and θ to get the exponential form $z = re^{i\theta}$. Now solve for w to get $z = e^w$.

4. Here is another way to see DeMoivre's Theorem in action. Recall from *Graphics* the rotation matrix

$$R = \begin{bmatrix} \cos(\theta) & -\sin(\theta) \\ \sin(\theta) & \cos(\theta)) \end{bmatrix}$$

Given a vector x, then Rx is the result of rotating x through the angle θ. Try this on $v = [1; 0]$ and use `vecplot` (see page 43 of *Graphics*) to plot $[v, Rv, R^2v, R^3v, \ldots, R^5v]$ for $\theta = 2*pi/5$. Now hold this plot with `hold`. Compare the results to plotting $[z, z^2, z^3, z^4, z^5, z^6]$ where $z = \cos(\theta) + i * \sin(\theta)$.

5. You can use `polyval` to evaluate $f(z)$ for any polynomial and any complex z. Let `r=(1:5) + i*(1:5)` and `f=expan(r)` (see page 38.) Now try `polyval(f,1+i)`. You should know the answer before you ask MATLAB. Some savings on complex operations can be obtained when evaluating real polynomials at a complex number. Suppose that we want to evaluate $f(x)$ at $z = a + b * i$ where $f(x)$ is a real polynomial. Let $g(x) = (x - z)(x - \bar{z}) = x^2 - 2ax + (a^2 + b^2)$. Use the Division Algorithm (deconv from page 37 in *Polynomials*) to get $f(x) = g(x)q(x) + r(x)$, then $f(z) = r(z)$ because $g(z) = 0$. Since $r(x)$ is a linear polynomial, there are very few complex operations to compute $r(z)$. Try this on f=ones(1,5) at z=1+2*i. Compare your answer with polyval(f,z).

6. Let f=[1,rand(1,5)] and r=roots(f). Compare f with expan(r). Do you know why this works? Now let f=rand(1,5) and try the same experiment. Can you explain the difference? How can you alter f so that f=expan(r) without changing the roots?

7. Make mesh graphs on the domain $z = x + iy$ where $-2 \le x \le 2$ and $-2 \le y \le 2$

(1) $\log(z)$
(2) $\log(\log(z))$
(3) $\cos(z)\sin(z)$
(4) $\sqrt{1 + z} + \sqrt{1 - z}$
(5) z^2
(6) z^3
(7) $z + 1/z$

Do both the real and the complex parts.

Errors

ABSTRACT

This is an introduction to floating point numbers and the kinds of errors that can occur in floating point computation. WARNING: This project contains explicit material which may be shocking.

MATLAB COMMANDS

    format, *, ^, \ , sqrt, roots, plot,

LINEAR ALGEBRA CONCEPTS

Conditioning

BACKGROUND

Computer arithmetic is a mine field of possible mistakes. In this project we are concerned with the way numbers are represented and manipulated by the machine. By understanding the basics of computer arithmetic we hope to be in a position to control or at least work around the sources of error. Just to set you up for the kinds of problems we will find, type

    10 ^ 20+200-10 ^ 20

I hope that you are not satisfied with the answer which was returned. The answer is quite different when you type

    10^ 20-10^ 20+200

The machine is not totally stupid. As you know from the *MATLAB Tutorial,* MATLAB has several display formats: `format short`, which is the default format, `format long`, `format short e`, and `format long e`. These are not to be confused with different **precision modes** in other languages. MATLAB always works in double precision. You can change the display format to suit your needs at any time. For the remainder of this discussion you should switch to

    format long e

Just to get a view of the format, type

    pi*10^ 179

Floating point numbers are represented in the form $m \times b^e$. Here m is called the **mantissa,** b is the **base,** and e is the **exponent.** The mantissa is a t digit number $m = m_1.m_2 \ldots m_t$ where $0 \le m_i < b$. The exponent e ranges inside $L \le e \le U$ where L is a negative integer and U is a positive integer. All of these parameters are machine dependent. The `format long e` display shows $t = 16$ and $b = 10$. You can discover U by the following short program

```
x=10; y=100; U=0;
while x~=y,
    x=y;
    y=10*y;
    U=U+1;
end
U
```

Similarly, L can be determined with

```
x=1; y=.1; L=1;
while x~=y,
    x=y;
    y=y/10;
    L=L-1;
end
L
```

The outcome of these programs varies greatly from machine to machine and the programs do not always run correctly. I encourage you to explore the sequence of values for x to get a better idea of the overflow and underflow limits.

Floating point numbers are an awkward number system. There are 10^{16} possible mantissas, thus in each interval $[10^e, 10^{e+1}]$ there are 10^{16} floating point numbers. But the interval has length $10^{e+1} - 10^e = 9*10^e$ which is 10 times longer than the previous interval $[10^{e-1}, 10^e]$ which has length $9*10^{e-1}$, and 10 times shorter than the next interval $[10^{e+1}, 10^{e+2}]$ which has length $9*10^{e+2}$. Thus the floating point numbers are arranged very densely near 0 and very thinly near 10^U.

Unfortunately, the machine does not work with complete accuracy on all of the floating point numbers which can be represented. The smallest number, x, with the property that $1 + x > 1$ is called the ***machine precision***. MATLAB claims this number is represented by the variable eps. We see that eps is close to 10^{-16}. Notice that .1*eps can be represented in the machine,

```
x=.1*eps
```

```
x==0
```

We still get the ***round-off error***

```
x=.1*eps+1
```

```
x==1
```

We also get round-off from large numbers (as you saw at the start of the discussion).

```
x=1/(.1*eps)
```

```
x+1==x
```

Round-off error is determined by the mantissa. ***Overflow*** occurs when a number x is created with $|x| > b^U$. Form the expressions 1e+U and 1e+U+1 by providing the value of U found in the little program listed above.

Underflow occurs when a number x is created with $0 < |x| < b^L$. Form the expressions 1e+L and 1e+(L-1) by inserting the value of L found above.

Generally, you can expect erratic results when working near L and U. Overflow and underflow are disasterous to a computation.

We can measure error in two ways. Suppose that x_{true} is the correct outcome and x_{comp} is the computed outcome. The ***absolute error*** is

$$|x_{true} - x_{comp}|$$

while the ***relative error*** is

$$\frac{|x_{true} - x_{comp}|}{|x_{true}|}.$$

The more important concept is relative error. If we ignore absolute values

$$x_{comp} = x_{true}(1 - \varepsilon)$$

where ε is the relative error.

Let's look at the effect of round-off error on arithmetic operations. For multiplication the mantissa is essentially doubled in size before rounding, thus $x_{true} = .m_1 \ldots m_t \ldots m_{2t} \times 10^e$ while $x_{comp} = .m_1 \ldots m_t \times 10^e$. The relative error is

$$\frac{x_{true} - x_{comp}}{x_{true}} = \frac{.m_{t+1} \ldots m_{2t} \times 10^{e-t}}{.m_1 \ldots m_{2t} \times 10^e} = \frac{.m_{t+1} \ldots m_{2t}}{.m_1 \ldots m_{2t}} \times 10^{-t}$$

Thus the round-off for multiplication is roughly machine precision. The same is true for division. Addition and subtraction are somewhat different. When we subtract two numbers which are close together, we invite the possibility of round-off. Look at

```
x=.1*eps
(x+1)-1
```

After what you have seen so far, you should not be surprised at the answer. Here $x_{comp} = 0$ while $x_{true} = x = .1 * eps$ and thus the relative error is

$$\frac{x_{true} - x_{comp}}{x_{true}} = \frac{x}{x} = 1.$$

That is, the relative error is 100%. Since the absolute error is only $.1 * eps$, this seems rather insignificant, but 10 of these will make an error of size eps, 100 an error of size $10 * eps$, and so on. Before you know it a significant error is present. Subtraction of numbers which are close together is a fundamental source of round-off error.

While MATLAB continually works in double precision (essentially 16 places), many times considerations of processing speed and the fact that your input data may only be accurate to 8 (or fewer) places, require you to use the single-precision mode in a compiled language like FORTRAN or C. These languages support both precision modes. Errors will accumulate more quickly in single precision.

Conditioning

Consider the system of equations

$$x + y = 1$$
$$x + (1 + eps)y = 1$$

Obviously, the solution to this system is $x = 1$, $y = 0$. Now suppose that through round-off or possibly a data collection error, the system is slightly perturbed to

$$x + y = 1$$
$$x + (1 + eps)y = 1 + eps$$

Here the solution is $x = 0$, $y = 1$ which is a large difference for such a minor change. When slight changes in the input data of a problem make dramatic changes in the output data, then we say that the problem is **ill conditioned.** Later, we will attempt to measure the conditioning of a problem by attaching a "condition number" to it. Conditioning is a property of the the problem, not the method used on the problem.

The conditioning problem immediately suggests the issue of testing a computed solution for accuracy. Suppose that in solving the matrix equation $Ax = b$ we get the computed solution x_c. We learned in grade school to check solutions by plugging the answer back into the equation. This amounts to checking if $r = Ax_c - b$ is close to $\vec{0}$. The vector $r = Ax - b$ is called the **residual.** Unfortunately, the residual is a poor indicator of accuracy. In the example above

```
A=[1, 1; 1, 1+eps]
b=[1; 1]
```

As before the true solution is `[1; 0]`. Now look at the residual of `z=[0; 1]`.

```
z=[0;1]

A*z-b
```

Normally this is a tolerable error, but look how far away z is from the solution! See *Norms and Condition Numbers* for ways of estimating error.

Stability

Some methods for solving a problem are better than others. When a method leads to serious round-off error, we say that the method is **unstable.** Here is an example.

Let $f(x) = \dfrac{1}{\sqrt{x^2+1}+1}$. We know that $\lim_{x\to 0} f(x) = .5$. Try the following program:

```
for i=5:10,
   x=10^ (-i);
   1/(sqrt(x^ 2+1)+1),
end
```

Now rationalize the denominator to get $\dfrac{\sqrt{x^2+1}-1}{x^2}$ and compute

```
for i=5:10,
   x=10^ (-i);
   (sqrt(x^ 2+1)-1)/x^ 2,
end
```

Can you explain the difference? The second method is *unstable.* An unstable method can lead a calculation into catastrophic round-off, as we see above. We will try to detect and avoid unstable methods.

PROBLEMS

1. Consider the sequence $x_0 = 1$, $x_1 = \dfrac{1-\sqrt{5}}{2}$ and $x_{n+1} = x_n + x_{n-1}$ for $n \geq 1$. Use MATLAB to compute x_{400}. You should have a large value for $|x_{400}|$. Now notice that x_1 satisfies the equation $x^2 - x - 1 = 0$, that is $x_1^2 = x_1 + 1$. Thus $x_2 = x_0 + x_1 = x_1^2$, $x_3 = x_1 + x_2 = x_1 + x_1^2 = x_1(1 + x_1) = x_1^3, \ldots, x_n = x_1^n$. Since $|x_1| < 1$, $\lim_{n\to\infty} x_n = \lim_{n\to\infty} x_1^n = 0$. Now compute $x_{400} = x_1^{400}$.

2. Checking for equality with floating point numbers is always dangerous. Try the following little program: (c is included to stop the iterations)

```
s=0; c=1; x=1/6;
while sv=1 & c<10,
    s=s+x, c=c+1;
end
```

You see the problem. The way around this problem is to use a new parameter tol. The value of tol will vary with each particular algorithm. In this example tol=eps works.

Now try

```
s=0; c=1; x=1/6;
while abs(s-1)>tol & c<10,
    s=s+x, c=c+1;
end
```

3. Round-off occurs with the dot product, and hence with matrix multiplication. Run the following little program:

```
for i=0:12,
    a=[a,10^ (12-i)+10^ (-i)];
end
a(14)=-sum(a)
```

With a value for a now let

```
    b=ones(size(a));
```

We are now ready to compute the dot product.

```
    b*a'
```

We know what the dot product should be,

```
    sum(a)
```

4. The Hilbert matrix provides an excellent illustration of the concept of ill conditioning. Try the following experiment:

```
    A=hilb(12);
    x=ones(12,1); b=A*x;
```

You should know the solution to Ax=b. Now form

```
    y=A\b
```

To get some idea of how far this is from the answer

```
    max(abs(x-y))
```

5. Quadratic equations provide a view of both conditioning and stability. If you have worked through *Polynomials*, then you have seen the conditioning problem for finding roots of a polynomial in problem 7 of *Polynomials*. Look at the equations

$$x^2 + 2x + 1 = 0$$
$$x^2 + 2x + 1.001 = 0$$
$$x^2 + 2.001x + 1 = 0$$

Try `roots` on each of these and notice how the roots are moved by these minor perturbations of the coefficients.

Now look at the standard way of computing square roots using the quadratic formula. Unless used properly, the quadratic formula is not stable. For the quadratic equation $ax^2 + bx + c = 0$

$$x = \frac{-b \pm \sqrt{b^2 - 4ac}}{2a}$$

Now try $x^2 + 10^9 x + 1$. Good approximations to the roots are given by $x = -10^{-9}, -10^9 + 10^{-9}$. Compute the relative error of your computed answers. Is it acceptable? The problem occurs when subtracting $-b + \sqrt{b^2 - 4ac}$ since the two terms are nearly equal. By rationalizing the numerator of

$$x = \frac{-b + \sqrt{b^2 - 4ac}}{2a}$$

we get a different expression for the same root

$$\frac{-2c}{b + \sqrt{b^2 - 4ac}}$$

Try this. Is it satisfactory?

6. To get a view of the distribution of the the floating point numbers plot the 1-place ($t = 1$) floating point numbers for $b = 10$ and $0 \le e \le 3$.

```
x=1:9; x=[x,10*x,100*x];
y=ones(size(x))
plot(x,y,'.')
```

To get a different view of this try

```
semilogx(x,y,'.')
```

7. Let $f(x) = x^8 - 8x^7 + 28x^6 - 56x^5 + 70x^4 - 56x^3 + 28x^2 - 8x + 1$. Use `polyplot` (see page 42 of *Graphics*) to plot $f(x)$ on the interval [.9, 1.1]. Now zoom in to [.99, 1.01] and [.999, 1.001]. What you are seeing is round-off error. Now you can see why some people refer to it as "round-off noise." Try increasing the resolution from the default value of 50 set in `polyplot` to 500.

COMMENTS

For a discussion of methods of detecting accuracy see *Norms and Condition Numbers.*

Chaos and Fractals

ABSTRACT

This project is a listing of a few MATLAB programs which illustrate some of the graphs associated with chaos and fractals. It is included in this book primarily for fun, though some of the data generated by these functions can be used in other projects.

MATLAB COMMANDS

```
plot
```

BACKGROUND

The Chaos Game

The program, srpnski(m,dist,n), gets its name from the mathematician W. Sierpinski. The only parameter that must be specified is m which determines the number of vertices that will be part of a regular polygon. For larger values of m it produces a graph which is similar to a snowflake. The program starts with a randomly chosen seed position given by the internal variable, s. At each stage one of the vertices is chosen at random and a new point is produced with is dist of the way from the old point to the vertex. The value of dist should be between 0 and 1. The default value is .5. This process is repeated n times. The default value of big is 1500.

```
function srpnski(m,dist,n)

% srpnski(m,dist,n). This creates a snowflake
% from "m" vertices using "n" iterations.
% The default value of n is 1500. "dist" is the
% value used to deterimine the distance from the
% previous point to a vertex. The default value
% is .5.

if nargin < 3, n=1500; end
if nargin < 2, dist=.5; end
clg
axis([-1,1,-1,1])
p=exp(2*pi*i*(1:m)/m);
plot(p,'*')
hold
s=rand+i*rand;
for j=1:n
   r=ceil(m*rand);
   s=dist*s+(1-dist)*p(r);
   plot(s,'.')
end
```

Try this out a few times for fun. Start with

```
srpnski(3)
srpnski(3,.5,2500)
srpnski(3,.5,5000)
```

This pattern is very clear now. Now try adjusting the dist parameter.

```
srpnski(3,.4)
srpnski(3,.2)
```

Try increasing the number of vertices.

```
srpnski(5)
```

The problem is that the dist parameter is not correct. Try something smaller.

```
srpnski(5,.4)
srpnski(5,.3)
```

The Barnsley Fern

The following program is the famous Barnsley Fern. The only external parameter in this program is n, the number of interations.

```
function fern(n)

% fern(n). This creates Barnsley's famous
% sprigwort fern using n iterations.

A1=[.85,.04;-.04,.85]; A2=[-.15,.28;.26,.24];
A3=[.2,-.26;.23,.22]; A4=[0,0;0,.16];
T1=[0;1.6]; T2=[0;.44];
T3=[0;1.6]; T4=[0;0];
P1=.85; P2=.07;
P3=.07; P4=.01;

clg
s=rand(2,1);
plot(s(1),s(2),'.')
hold
for j=1:n
   r=rand;
   if r <= P1, s=A1*s+T1;
   elseif r <= P1+P2, s=A2*s+T2;
   elseif r <= P1+P2+P3, s=A3*s+T3;
   else s=A4*s;
   end
   plot(s(1),s(2),'.')
end
```

Try the following to see the shape of the fern emerge.

```
fern(100)
fern(500)
fern(1000)
fern(3000)
fern(5000)
```

You will notice that the program has internal variables A1, A2, A3, A4, T1, T2, T3, T4, P1, P2, P3, P4. In the exercises there are suggestions for changing these values to produce different pictures.

The Logistic Equation

The logistic equation now appears in many books on differential equations. The following program is based on the iteration $x_{new} = cx_{old}(1 - x_{old})$. This is a simple model of population growth. In this model x_{old} would be the percentage of a population at a particular generation measured against a maximum size for the population.

```
function y=logistic(c,ax,n)

% logistic(c,ax,n). Performs n iterations
% of the logistic iteration xnew=c*xold*(1-xold)
% and plots. Default value of n is 50. c is
% a damping constant. The seed position is
% randomly set. ax sets the axis. The default
% setting of ax is [0,n,0,1].

if nargin < 3, n=50; end
if nargin < 2, ax=[0,n,0,1]; end

x=rand;
for i=1:n,
    x=c*x*(1-x);
    y=[y,x];
end

clg
axis(ax); hold
plot(1:n,y)
grid;
```

Try these values.

```
logistic(2.8)
logistic(2.2)
logistic(1.6)
logistic(1.3)
```

See problem 4 for some other values to try with `logistic`.

The Hénon Map

The **Hénon map** produces one of the most cited examples of a **strange attractor.** It consists of the following simple iteration:

$$x_{new} = 1 + y_{old} - 1.4x_{old}^2$$
$$y_{new} = .3y_{old}$$

The function `strange(n,v)` shows how this iteration accumulates. The set of points which emerges from applying an function to a single point is called the **orbit** of the starting point. The orbits are extremely sensitive to the input point `v`. With the Hénon map an orbit will either be drawn to the attractor or the orbit will go to infinity.

```
function strange(n,v)

% strange(n,v). Plots the orbits of the
% henon strange attractor. n is the number
% of iterations, the default value is 100.
% v is the input point, the default is (0,0)

if nargin < 2, v=[0;0]; end
if nargin < 1, n=100; end

clg
plot(v(1),v(2),'.');
grid
axis([-1.5, 1.5, -.5, .5]);
hold

for i=1:n,
    v=[v(2)+1-1.4*v(1)^2;  .3*v(1)];
    plot(v(1),v(2),'.')
end
```

Try a few runs:

```
strange
strange(500)
strange(1500)
```

```
strange(5000)
```

Now zoom in using the `axis` command.

```
ax=[1.2,1.3,-.01,.01];
```

```
axis(ax)
```

See problem 5 for more `strange` things to do.

PROBLEMS

1. Try the following values in the `fern.m` program. Call the new program `srptri.m`. It should be familiar.

```
A1=[.5,0;0,.5];  A2=[.5,0;0,.5];
A3=[.5,0;0,.5];
T1=[1;1];  T2=[1;50];  T3=[50;50];
P1=.33;  P2=.33;  P3=.34;
```

2. Try the following values in the `fern.m` program. Call the new program `srptree.m`.

```
A1=[0,0;0,.5];  A2=[.42,-.42;.42,.42];
A3=[.42,.42;-.42,.42];  A4=[.1,0;0,.1];
T1=[0;0];  T2=[0;.2];
T3=[0;.2];  T4=[0;.2];
P1=.05;  P2=.4;
P3=.4;  P4=.15;
```

3. Try the following values in the `fern.m` program. Call the new program `srpsq.m`. The plot here will surprise you.

```
A1=[.5,0;0,.5];  A2=[.5,0;0,.5];
A3=[.5,0;0,.5];  A4=[.5,0;0,.5];
T1=[1;1];  T2=[1;50];
T3=[50;1];  T4=[50;50];
P1=.25;  P2=.25;
P3=.25;  P4=.25;
```

4. Study some of the other values of `logistic`. Try these:

```
logistic(4), logistic(3.5), logistic(3.2), logistic(3)
```

Work your way slowly towards `logistic(1)`. Make axis adjustment in your plots to get a better view of what is happening. Go on to see what happens below $c = 1$.

```
ax=[0,50,0,.2]
logistic(1,ax), logistic(.9,ax), logistic(.8,ax), logistic(.5,ax)
```

5. Try some more iterations of the Hénon map using different starting points: (-1,0), (-.5,0), (-.5,.5), (0,.5). Use the `axis` command to zoom in on different locations of the attractor.

6. Change the assignment line in `strange.m` from

```
v=[v(2)+1-1.4*v(1)^ 2;  .3*v(1)];
```

to

```
v=[1-v(2)+abs(v(1));v(1)];
```

Change the default value of v to v=[-.1;0] and remove the axis setting. Call this new program ginger.m. Try the following:

```
ginger
ginger(1000)
ginger(2500)
```

Use the axis command to zoom in on some of the clusers of points.

7. The following iteration is also associated with a study by Hénon. The program henon listed below will plot n points of the orbit of the starting point v. Try some of the following experiments with it.

```
function henon(r,v,n)

% henon(r,v,n). Plots the orbits of the
% henon iteration. r is the angle,
% n is the number of iterations,
% default=100; v is the starting point.

if nargin < 3, n=1000; end
if nargin < 2, v=[.1;.1]; end
if nargin < 1, r=pi/2; end

clg
hold

cs=cos(r);
sn=sin(r);
for i=1:n,
  v=[v(1)*cs-(v(2)-v(1).^2)*sn;
     v(1)*sn+(v(2)-v(1).^2)*cs];
  plot(v(1),v(2),'.')
end
```

Try the following:

```
henon
henon(.6*pi)
henon(.503*pi)
henon(.501*pi)
```

You might want to study the situation around $\pi/2$ more closely to see how the graph changes. Now try moving the point v a little but keeping r=.5*pi.

```
henon(.5*pi,[.5;.5])
```

```
henon(.5*pi,[.5;.5],3000)
```

You should see some lace appearing. Use `axis` to zoom in on one of those little ovals.

```
ax=[.4 .6 .4 .6]
```

```
axis(ax)
```

Now nudge the point slightly and then zoom in.

```
henon(.5*pi,[.51;.51])
```

```
axis(ax)
```

Flops

ABSTRACT

To measure the efficiency of an algorithm we count floating point operations or flops. In this project we define "flops" and go over some basic techniques for estimating flops.

MATLAB COMMANDS

```
clock, etime, flops, rref, +, *, \
```

LINEAR ALGEBRA CONCEPTS

Gaussian Elimination

BACKGROUND

We are going to consider two ways of measuring the amount of work or running time of an algorithm. The most obvious is to measure the time using a clock. MATLAB provides a way to access the clock in your machine with the commands `clock`, which returns the machine time, and `etime`, which will measure the elapsed time. Suppose that you want to time a program

```
a=rand(10); b=rand(10,1)

t=clock; a\b; etime(clock,t)
```

Try repeating the same line several times. You will most likely get some variation in the time. Try repeating the operation 20 times and averaging.

```
n=20;

t=clock; for i=1:n, a\b; end, etime(clock,t)/n
```

If you repeat this several times you should notice less variation in the answer. Try it for other values of `n`. When you are working in MATLAB this is a very useful way to measure the speed of a program.

To get some measure of efficiency which is independent of the machine or the language that it is written in, you should count the operations performed by the algorithm. A *flop* or *floating point operation* is either an addition, subtraction, multiplication, or division operation. In floating point arithmetic these operations are counted as one *flop*. In complex number arithmetic, addition and subtraction are counted as 2 flops; multiplication and division are counted as 6 flops. We are going to concentrate on real arithmetic. These are the MATLAB conventions for counting flops, there are other conventions. For example, one convention counts the two operations of an addition and a multiplication together as a flop. Other operations such as `sin`, `log`, etc. are also counted as flops by MATLAB. MATLAB keeps a running flop counter throughout a session. In MATLAB the command `flops` will give a count of the number of flops executed since the last time the counter was reset. The command `flops(0)` will reset the flop counter. Try

```
flops(0); for i=1:20, a\b; end, flops
```

The dot product of vectors $x = (x_1, \ldots, x_n)$ and $y = (y_1, \ldots, x_n)$ is

$$x \cdot y = x_1 y_1 + \cdots + x_n y_n$$

has n multiplications and $n - 1$ additions for a total of $2n - 1$ flops. MATLAB counts this as $2n$ flops for easier calculation. This count varies a bit with different implementations.

Before we consider operations like addition and multiplication, let's look at a general principle. Suppose we have a loop

$$\text{for i=1:k, P(i), end}$$

where each execution of the program P(i) requires $F(i)$ flops. Then the loop requires

$$\sum_{i=1}^{k} F(i) \quad \text{flops}$$

Now look at the program for addition

```
for i=1:m,
    for j=1:n,
        c(i,j)=a(i,j)+b(i,j);
    end
end
```

According to the method, the flops are counted as

$$\sum_{i=1}^{m}\sum_{j=1}^{n} 1 = \sum_{i=1}^{m} n = mn.$$

Let's set this up with an example. Suppose we let

$$\text{a=rand(8); b=rand(8);}$$

Now if we apply the formula to the matrices a and b in the example above where $m = n = 8$ we should get 64 flops. You can see how this works

$$\text{flops(0), a+b; flops}$$

Did the answer coincide with our calculation? Notice that for $n \times n$ matrices the flop count for addition is n^2. If you created the file add.m (see page 18 in the *MATLAB Tutorial,*) then you can try

$$\text{flops(0), add(a,b); flops}$$

Now try the clock time

$$\text{t=clock; a+b; etime(clock,t)}$$

$$\text{t=clock; add(a,b); etime(clock,t)}$$

This is one of the essential lessons about using MATLAB. When you can, use the built-in functions rather than an m-file. They save programming time, run faster, and are usually more accurate. It is not unusual when comparing a MATLAB built-in function to an m-file for the built-in function to have a larger flop count but a smaller clock time.

We are now going to estimate the flops for Gaussian Elimination. We begin by considering the elementary row operations. The flop counts on the elementary row operations on a matrix with n columns are easily computed. See page 25 of *Systems of Linear Equations* for the elementary row operations.

Operation	Flops
elerow1(A, i, j)	0 flops
elerow2(A, r, i)	n flops
elerow3(A, r, i, j)	$2n$ flops

The following function is called badgauss. It is intended for illustrative purposes only.

```
function A=badgauss(A)

% A=badgauss(A). Given a square
% matrix A badgauss returns a
% matrix in triangular form. It
% does no row switching.

[m,n]=size(A);
for j=1:n-1,
 for i=j+1:n,
  A=elerow3(A,-A(i,j)/A(j,j),j,i);
 end
end
```

To count the flops we notice that the elerow3's cost $2n$ flops so that we get

$$\sum_{j=1}^{n-1} \sum_{i=j+1}^{n} 2n = 2n \sum_{j=1}^{n-1} (n-j)$$

$$= 2n \left(n(n-1) - \frac{n(n-1)}{2} \right)$$

$$= n^3 - n^2$$

It may come as a surprise, but this is comparable to matrix multiplication for square matrices, see problem 1.

We will give a short example of how to estimate flops for a recursive program. Consider the following program for computing 2^n (page 20 in the *MATLAB Tutorial*.)

```
if n==0, y=1;
    else y=2*twoexp(n-1);
end
```

The first step is to find a recurrence relation which counts the flops. Let $F(n)$ represent the flops for computing `twoexp(n)`, then we see that $F(0) = 0$ and $F(k+1) = F(k) + 1$, the +1 comes from the multiplication by 2 and the $F(k)$ comes from the recursive call to `twoexp(n-1)`. In this example we can evaluate $F(n)$ easily

$$F(n) = F(n-1) + 1 = F(n-2) + 2 = \ldots = F(n-n) + n = n.$$

This recurrence is particularly easy to evaluate It is usually much more difficult to find an expression for a recurrence relation.

PROBLEMS

1. The following is the program for matrix multiplication which was given on page 19 in the *MATLAB Tutorial*.

```
for i=1:m,
  for j=1:l,
    for p=1:n,
      c(i,j)=c(i,j)+a(i,p)*b(p,j);
    end
  end
end
```

Count the number of flops to compute the product of an $m \times n$ matrix and a $n \times l$ matrix. Let A=rand(4,5) and B=rand(5,3). Use MATLAB's `flops` to find the number of flops for `mult(A,B)` and `A*B`. Now use `etime` to compare the clock times.

2. Let A=rand(5)+i*rand(5); and B=rand(5)+i*rand(5). First decide how many flops you think will be used in A+B, then check it using `flops`. How many flops should be used in A*B? Check it with `flops`.

3. MATLAB's `A\b` is a sophisticated function which uses different methods depending on the size of A. Type `help slash` for the discussion on this. Let A=rand(10) and b=rand(10,1) and compare the flops between `A\b` and `rref([A,b])`. Now let A=rand(10,5) and compare the flops between `A\b` and `rref([a,b])`.

4. Count the flops for `backsub(A,b)` (see page 91 of *Triangular Systems*) for an $n \times n$ matrix A. Compare your predicted results with the actual flops used on A=triu(rand(10)) and b=rand(10,1). Variations in programming style may cause discrepancies in the actual flops counts.

5. Count the flops for `trisolve(A,b)`, the function which computes the solution to Ax=b where A is a tridiagonal matrix. See page 92 of *Triangular Systems* for `trisolve`. Compare your predicted results with the actual flops used on A=triu(tril(rand(10),1),-1) and b=rand(10,1).

6. The classic method of synthetic division for evaluating a polynomial represented by f at the value x is given by the following little program:

```
n=length(f);
[q,r]=deconv(f,[1,-x]);
r(n);
```

How does this compare with `polyval` (see page 36) for evaluating the polynomial given by `f=1:20` at `x=2` for the number of flops used?

7. In Exercise 11 of *Polynomials* you were asked to write a function `horner` which used the MATLAB function \ to evaluate a polynomial. Your `horner` should be quite fast if you formed the matrix using

```
A=-t*diag(ones(1,n-1),-1)+eye(n)
```

and used MATLAB's \ Compare `horner` and `polyval` for flops and clock time on the polynomial `f=1:20` and `x=2`. If you do this several times you may notice that the clock time is fairly erratic. A way to get a reasonable answer is to average a number of runs as follows:

```
t=clock;
for i=1:20, polyval(f,x); end
etime(clock,t)/20
```

8. Which is better from a flops point of view `A\b` or `inv(A)*b`? Devise an experiment to decide.

9. Given two $n \times n$ matrices A and B and an $n \times 1$ vector v, there are two ways to parenthesize the product ABv, namely, $(AB)v$ and $A(Bv)$. Run some experiments using random 10×10 matrices to determine if there is any difference in the flop counts. Which is better? If you just give MATLAB `A*B*v`, what does it do?

COMMENTS

Here are some standard formulas which are useful for counting flops.

$$\sum_{i=1}^{n} i = \frac{n(n+1)}{2}$$

$$\sum_{i=1}^{n} i^2 = \frac{n(n+1)(2n+1)}{6}$$

Norms and Condition Numbers

ABSTRACT

We are interested in determining the reliability of a solution. The condition number is a way of estimating the accuracy of a computed solution.

MATLAB COMMANDS

    norm, cond, plot, mesh, max, sqrt, flops, inv

LINEAR ALGEBRA CONCEPTS

Norm, Condition Number

BACKGROUND

We have already mentioned the **2-norm** also called the ***Euclidean Norm*** (see page 6) of a vector which was defined to be

$$\|x\|_2 = \left(\sum_{i=1}^{n} x_i^2 \right)^{1/2}$$

There are other useful norms: The **1-norm**

$$\|x\|_1 = \sum_{i=1}^{n} |x_i|$$

and the ∞-**norm**

$$\|x\|_\infty = \max_{1 \leq i \leq n} |x_i|$$

We have the following relationships:

$$\|x\|_1 \geq \|x\|_2 \geq \|x\|_\infty$$

MATLAB has a single function norm which computes all of these. In MATLAB try

    x=1:5; norm(x,1), norm(x,2), norm(x,inf)

A **norm** is defined to satisfy these properties:

    (1) $\|\vec{0}\| = 0$
    (2) if $x \neq \vec{0}$, then $\|x\| > 0$
    (3) $\|rx\| = |r| \|x\|$ for all scalars r.
    (4) $\|x + y\| \leq \|x\| + \|y\|$.

Property (4) is called the ***triangle inequality.***

We are interested in getting versions of these norms for matrices. That is, we want to define $\|A\|$ for an arbitrary matrix. The definition will satisfy the same properties as the vector norms, *viz.*

    (1) $\|0\| = 0$
    (2) if $A \neq \vec{0}$, then $\|A\| > 0$
    (3) $\|rA\| = |r| \cdot \|A\|$ for all scalars r.
    (4) $\|A + B\| \leq \|A\| + \|B\|$.

In addition we want the **consistency property,**

$$\|AB\| \leq \|A\| \cdot \|B\|.$$

In order to get these properties we define

$$\|A\| = \max_{\|x\|=1} \|Ax\|.$$

This is the largest value of $\|Ax\|$ on the "unit circle," where $\|x\| = 1$.

Theorem 1. *If $\|A\|$ is defined as above, then properties (1), (2), (3), and (4) and the consistency property are satisfied.*

The 2-norm is the most important but it is difficult to calculate. MATLAB offers a function `norm` to compute these norms. In view of the following facts,

$$\|A\|_1 = \max_{1 \leq j \leq n} \left(\sum_{i=1}^{n} |a_{ij}| \right) = \text{norm}(A, 1)$$

$$\|A\|_\infty = \max_{1 \leq i \leq n} \left(\sum_{j=1}^{n} |a_{ij}| \right) = \text{norm}(A, \text{inf})$$

we see that the 1-norm and the ∞-norm are easily computed. Another norm on matrices which is easily computed is the **Frobenius norm.**

$$\|A\|_F = \left(\sum_{i=1}^{n} \sum_{j=1}^{n} a_{ij}^2 \right)^{1/2}$$

In MATLAB $\|A\|_F = $`norm(A,'fro')`. The 2-norm can also be computed using
`norm(A,2)`.

See *The Singular Value Decomposition* to see how MATLAB computes the 2-norm.

The norm of a matrix can be useful in getting quick bounds on a problem. For example, in *Eigenvalues* we show that $|\lambda| \leq \|A\|$ for all eigenvalues λ of A.

The Condition Number for Solving Ax=b

We are now in a position to define the condition number for solving $Ax = b$ for an invertible matrix A. If $\|A\|$ is any norm, then the **condition number of A with respect to** $\|A\|$ is

$$\text{cond}(A) = \|A\|\|A^{-1}\|.$$

The MATLAB function `cond(A)` computes the condition number for the 2-norm. The importance of the condition number is found in the next theorem. Our goal is to get an estimate on the relative error which occurs in computing a solution to the equation $Ax = b$.

Theorem 2. *Let A be an invertible matrix and cond(A) be the condition number for some norm $\|A\|$ satisfying the consistency property. Suppose that x_t is the true solution to $Ax = b$, x_c is the computed solution, and $r = Ax_c - b$ is the residual. Then*

$$\frac{\|x_c - x_t\|}{\|x_t\|} \leq cond(A)\frac{\|r\|}{\|b\|}$$

PROOF: $A(x_c - x_t) = Ax_c - Ax_t = Ax_c - b = r$, so that $x_c - x_t = A^{-1}r$ and by the consistency property $\|x_c - x_t\| \leq \|A^{-1}\|\|r\|$. This gives

$$\frac{\|x_c - x_t\|}{\|x_t\|} \leq \|A\|\|A^{-1}\|\frac{\|r\|}{\|x_t\|}\frac{1}{\|A\|}.$$

Since $Ax_t = b$ we have, by consistency, $\|b\| \leq \|A\| \cdot \|x_t\|$, hence $\frac{1}{\|A\|\|x_t\|} \leq \frac{1}{\|b\|}$ and

$$\frac{\|x_c - x_t\|}{\|x_t\|} \leq cond(A)\frac{\|r\|}{\|b\|}.$$

■

We would like to check our accuracy by computing the norm of the residual r. It is possible for the residual to be small while the error in the solution is fairly high. (See problem 3 in *Triangular Systems*) The theorem explains this by appealing to the size of the condition number, which must then be very large. If the condition number is small and the residual is small then we can be confident that the error in the solution is small. The rule of thumb about the condition number is that if cond(A) $\approx 10^n$, then at most n places of accuracy are lost. In MATLAB where 16 places are held, then we would expect $16 - n$ places of accuracy. This rule is based on the fact (which is not proven here) that when using Gaussian Elimination

$$\|x_c - x_t\| \approx cond(A)\|x_c\|/10^{16}$$

so that if cond(A) $\approx 10^n$,

$$\|x_c - x_t\| \approx \|x_c\|/10^{16-n}$$

In words, the error is occurring after the first $16 - n$ places, so that the first $16 - n$ places should be correct.

Notice that Theorem 2 is not specific about which condition number (norm) is being used; any norm which satisfies the consistency property works. The condition number is difficult to compute from it's definition, because it is necessary to compute A^{-1}, which is as hard as solving the original problem $Ax = b$. While MATLAB does support cond(A) which finds the condition number with respect to the 2-norm, this is done at some expense. Try

```
A=rand(10); flops(0); cond(A), flops
```

Thus MATLAB also supports rcond, an estimate for $1/cond(A)$ where cond(A) is the condition number with respect to the 1-norm. Try

```
A=rand(10); flops(0); rcond(A), flops
```

The Condition Number for Matrix Multiplication

It is possible to develop the notion of a condition number for virtually any problem. As a final example we are going to consider the condition number for matrix multiplication. We define this as

$$\text{cond}(A, B) = \frac{\|A\|\|B\|}{\|AB\|}$$

As with the condition number for solving $Ax = b$ which was only defined for invertible matrices A, this condition number is only going to be defined for matrices A and B where $AB \neq 0$. Unfortunately this circumstance can fail fairly frequently for matrices. For example

$$A = \begin{bmatrix} 1 & -1 \\ -1 & 1 \end{bmatrix} \qquad B = \begin{bmatrix} 1 & 0 \\ 1 & 0 \end{bmatrix}$$

Notice that $AB = 0$ but $BA \neq 0$. Thus $\text{cond}(A, B)$ is not defined but $\text{cond}(B, A)$ is defined. If A is a row vector and B is a column vector, then $\text{cond}(A, B)$ will be undefined when they are orthogonal.

Theorem 3. *Suppose that $C = AB$, A is moved by A_E, B by B_E and C by C_E to obtain new matrices $A+A_E$, $B+B_E$, and $C+C_E$, which also satisfy $C+C_E = (A+A_E)(B+B_E)$. Then*

$$\frac{\|C_E\|}{\|C\|} \leq \text{cond}(A, B) \left(\frac{\|B_E\|}{\|B\|} + \frac{\|A_E\|}{\|A\|} \right)$$

PROOF: (sketch) Expanding $C+C_E = (A+A_E)(B+B_E) = AB + AB_E + A_E B + A_E B_E)$, yields

$$C_E = AB_E + A_E B + A_E B_E$$

Now taking the norm we get

$$\|C_E\| \leq \|A\|\|B_E\| + \|A_E\|\|B\| + \|A_E\|\|B_E\|$$

and dividing by $\|C\|$ gives

$$\frac{\|C_E\|}{\|C\|} \leq \frac{\|A\|\|B\|}{\|AB\|} \left(\frac{\|A\|\|B_E\|}{\|A\|\|B\|} + \frac{\|A_E\|\|B\|}{\|A\|\|B\|} + \frac{\|A_E\|\|B_E\|}{\|A\|\|B\|} \right)$$

Since we will usually have $\|A_E\|\|B_E\|$ small enough to ignore,

$$\frac{\|C_E\|}{\|C\|} \leq \frac{\|A\|\|B\|}{\|AB\|} \left(\frac{\|B_E\|}{\|B\|} + \frac{\|A_E\|}{\|A\|} \right)$$

PROBLEMS

1. Let `A=diag([2,3,4,5])` and compute the 2-norm and the `cond(A)`. Can you tell what the 2-norm and the condition number with respect to the 2-norm will be for a diagonal matrix?

2. Use the MATLAB `norm` function to compute condition numbers with respect to the 1-norm, the 2-norm, the ∞-norm, and the Frobenius norm for `A=rand(10)`, `pascal(10)`, `vander(1:10)`, `hilb(10)`, and `hilb(5)^2`. Compare your answers obtained using the 1-norm and the 2-norm with MATLAB's `cond(A)` and `1/rcond(A)`.

3. Recall the formula

$$\frac{\|x_c - x_t\|}{\|x_t\|} \le \text{cond}(A)\frac{\|r\|}{\|b\|}$$

from Theorem 2. We can illustrate this theorem by creating some random noise to play the role of x_c. Let `A=rand(10)`, `b=A*ones(10,1)`, so that the solution to $Ax = b$ is `xt=ones(10,1)`. Let `noise=10^(-5)*rand(10,1)`, `xc=xt+noise`, and `r=A*xc-b`. How well does the residual r estimate the error, $\|xc - xt\|$? Does the rule of thumb about the number of places of accuracy determined by the condition number work here. Try this again for `noise=10^(-10)`. What happens when you take `A=hilb(10)`?

4. To get a view of the "unit disc," $\{(x, y) : \|(x, y)\| \le 1\}$, with respect to each of the norms, use MATLAB's mesh feature.

```
[x,y]=meshgrid(-2:.1:2,-2:.1:2);
r1=abs(x)+abs(y)<=1;
```

r1 will be 1 for those points inside the disc for the 1-norm and 0 for those points outside the disc. Now see what this looks like with `mesh`.

```
mesh(r1)
```

Now we can try the other norms.

```
r2=sqrt(x.^2+y.^2)<=1; mesh(r2)
rinf=max(abs(x),abs(y))<=1; mesh(rinf)
```

To see how these are related

```
mesh(r1+r2+rinf)
```

5. Recall the definition given for the matrix norms:

$$\max_{\|x\|=1}\|Ax\|$$

We are going to set up a graphic computation of the 2-norm of the matrix

```
A=[1, 2; 3, 3].
```

First set the axes

```
axis([-5,5,-5,5])
```

Now we set up the unit circle.

```
t=0:pi/20:2*pi; x=cos(t); y=sin(t);
plot(x,y), hold
```

```
A=[1,2;3,3]; z=A*[x;y];
x=z(1,:); y=z(2,:);
plot(x,y)
```

You can see how A has stretched the circle into an ellipse. The definition of the 2-norm tells us to find the maximum 2-norm of a point on this ellipse.

```
[norm2,i]=max(sqrt(x.^2+y.^2))
```

To compare this to MATLAB's 2-norm try

```
norm(A)
```

You can see the vector where the max occurs using `vecplot` (see page 43 of *Graphics*)

```
vecplot([x(i);y(i)]); hold
```

We can repeat this process for the 1-norm and the ∞-norm. The major differences are the shape of the unit circle and, of course, the definition of the norm. To do the 1-norm,

```
x=1:-.05:-1; y=-abs(x)+1;
x1=-1:.05:1; y1=abs(x1)-1;
x=[x,x1]; y=[y,y1];
plot(x,y)
```

You stretch the "unit circle" using A just as above and compute the norm by

```
[norm1,i]=max(abs(x)+abs(y)),
```

Again this can be plotted and viewed just as above. Use `norm(A,1)` to see MATLAB's answer. Do this for the ∞ -norm. What does the "unit circle" look like?

COMMENTS

MATLAB computes the 2-norm and the condition number for the 2-norm using the singular value decomposition. For more information on this topic see *The Singular Value Decomposition*.

Triangular Systems

ABSTRACT
We consider the problem of solving a matrix equation $Ax = b$ where A is either a triangular or a tridiagonal matrix. The methods of back substitution and forward substitution are introduced.

MATLAB COMMANDS
$*, -, \backslash$

LINEAR ALGEBRA CONCEPTS
Upper Triangular Matrix, Lower Triangular Matrix, Tridiagonal Matrix, Back Substitution, Forward Substitution, Elementary Row Operations

BACKGROUND

Upper Triangular Systems
We will start by considering an upper triangular system.

$$a_{11}x_1 + a_{12}x_2 + \ldots + a_{1n}x_n = b_1$$
$$a_{22}x_2 + \ldots + a_{2n}x_n = b_2$$
$$\vdots$$
$$a_{nn}x_n = b_n$$

This is called **upper triangular** since the only nonzero entries of the coefficient matrix lie on or above the main diagonal, that is $a_{ij} = 0$ if $i > j$. The coefficient matrix looks like

$$A = \begin{bmatrix} a_{11} & a_{12} & \cdots & a_{1n} \\ 0 & a_{22} & & a_{2n} \\ \vdots & \ddots & \ddots & \vdots \\ 0 & \cdots & 0 & a_{nn} \end{bmatrix}$$

The solution to the system can be found easily (assuming that the diagonal elements are nonzero!) using **back substitution** by starting at the bottom row. Thus

$$x_n = b_n/a_{nn}$$
$$x_{n-1} = (b_{n-1} - a_{n-1,n}x_n)/a_{n-1,n-1}$$

and, in general,

$$x_i = (b_i - \sum_{k=i+1}^{n} a_{ik}x_k)/a_{ii}$$

As each of $x_{i+1}, \ldots, x_n$ is known and $a_{ii} \neq 0$, this last formula makes sense. In MATLAB

```
A=triu(rand(4)); b=rand(4,1);
x=zeros(4,1);
x(4)=b(4)/A(4,4)
x(3)=(b(3)-A(3,4)*x(4))/A(3,3)
x(2)=(b(2)-A(2,:)*x)/A(2,2)
x(1)=(b(1)-A(1,:)*x)/A(1,1)
```

See if we're close

```
A*x-b
```

Lower Triangular Systems

A lower triangular system has the form

$$
\begin{aligned}
a_{11}x_1 & & = b_1 \\
a_{21}x_1 + a_{22}x_2 & & = b_2 \\
\vdots & & \vdots \\
a_{n1}x_1 + a_{n2}x_2 + \ldots + a_{nn}x_n & = b_n
\end{aligned}
$$

This type of system has a **lower triangular** coefficient matrix.

$$
A = \begin{bmatrix}
a_{11} & 0 & \ldots & 0 \\
a_{21} & a_{22} & \ddots & \vdots \\
\vdots & & \ddots & 0 \\
a_{n1} & a_{n2} & \ldots & a_{nn}
\end{bmatrix}
$$

The lower triangular systems can be solved using **forward substitution.**

$$
x_1 = b_1/a_{11}
$$
$$
x_2 = (b_2 - a_{21}x_1)/a_{22}
$$

with the general formula

$$
x_i = (b_i - \sum_{k=1}^{i-1} a_{ik}x_k)/a_{ii}.
$$

The ability to solve triangular systems is convenient, and enhanced by the fact that it is possible (with minor technical difficulties) to factor a matrix into a product of an upper triangular and a lower triangular matrix, called the LU Decomposition (see *The LU Decomposition* for some of the features of this factorization.) We will now look at how to use the factorization to solve general systems.

Suppose that we want to solve $Ax = b$. If we can write $A = LU$ where L is lower triangular and U is upper triangular, then first solve $Ly = b$ using forward substitution and then solve $Ux = y$ using back substitution. Putting this together

$$
Ax = LUx = Ly = b.
$$

This is an efficient and reasonably accurate method.

Tridiagonal Systems

A **tridiagonal matrix** has the form

$$\begin{bmatrix} a_{11} & a_{12} & 0 & \cdots & & \cdots & 0 \\ a_{21} & a_{22} & a_{23} & & \ddots & & \vdots \\ 0 & a_{32} & a_{33} & a_{34} & & \ddots & \vdots \\ \vdots & \ddots & \ddots & \ddots & & \ddots & 0 \\ \vdots & & \ddots & a_{n-1,n-2} & a_{n-1,n-1} & a_{n-1,n} \\ 0 & \cdots & \cdots & 0 & a_{n,n-1} & an, n \end{bmatrix}$$

Formally, $a_{ij} = 0$ if $j > i + 1$ or $j < i - 1$. Tridiagonal systems occur frequently in applications.

The way we approach solving $Ax = b$ is to use elementary row operations of type 3 to convert this into an upper triangular system. First form the matrix $B = [A, b]$. Now we want to wipe out a_{21} in B. This is accomplished by assigning

$$B = \text{elerow3}(B, -B(2, 1)/B(1, 1), 1, 2)$$

See page 25 for the definition of elerow3. Try this in MATLAB. You will need the function elerow3 (see page 30.)

```
**** A=triu(tril(rand(4),-1),1),
b=rand(4,1), B=[A,b]
B=elerow3(B,-B(2,1)/B(1,1),1,2)
B=elerow3(B,-B(3,2)/B(2,2),2,3)
B=elerow3(B,-B(4,3)/B(3,3),3,4)
```

This gives an upper triangular matrix. The back substitiution is very easy.

```
x=zeros(4,1); x(4)=B(4,5))/B(4,4),
x(3)=(B(3,5)-B(3,4)*x(4))/B(3,3)
x(2)=(B(2,5)-B(2,3)*x(3))/B(2,2)
x(1)=(B(1,5)-B(1,2)*x(2))/B(1,1)
```

Now check with A*x-b.

PROBLEMS

1. Let
```
A=[1,2,3,4;0,5,6,7;0,0,8,9;0,0,0,10]
b=ones(4,1).
```
Solve Ax=b using back substitution.

2. Write a MATLAB function x=backsub(A,b) which solves Ax=b by back substitution. *The program should not have any nested loops.* You should include an error message and a break to prevent division by 0. Check your backsub on the matrix in problem 1.

3. The following is a variation of an example due to Wilkinson. Let
```
W=rand(5)
```

```
W=diag(10^ -5*diag(W))+triu(W,1)
b=W*ones(5,1)
```

Now use your backsub (or do a back substitution as in problem 1) and MATLAB's W\b to find answers. How close are these to the correct answer? One way to measure this is

```
max(abs(ones(5,1)-x))
```

What is the size of the residual r=W*x-b? How accurate is the residual as a check for accuracy.

4. Devise a method similar to that of problem 1 to do forward substitutions. Try it on the matrix

```
A=[10,0,0,0;9,8,0,0;7,6,5,0;4,3,2,1]
b=ones(4,1).
```

5. Write a MATLAB function x=forsub(A,b) which accepts a lower triangular matrix A and solves Ax=b by forward substitution.

6. Starting with a random 5×5 lower triangular matrix L and a random 5×5 upper triangular matrix U, let A=L*U and b=A*ones(5,1). Now apply the method described in the Background discussion using forsub and backsub to solve Ax=b. Compare your answer with MATLAB's A\b.

7. Apply the method described in the Background discussion for solving a tridiagonal system to the system:

$$\begin{bmatrix} 2 & 3 & 0 \\ 4 & 5 & 6 \\ 0 & 7 & 8 \end{bmatrix} x = \begin{bmatrix} 1 \\ 2 \\ 3 \end{bmatrix}$$

Compare your solution with MATLAB's A\b.

8. Write a MATLAB function x=trisolve(A,b) which finds the solution to a tridiagonal system using the method described above. You should include an error message to prevent division by zero.

9. Try your trisolve to solve Cx=b where

```
C=4*eye(10)+diag(ones(1,9),1)
C=C+diag(ones(1,9),-1)
b=(1:10)'
```

Compare the solution with MATLAB's C\b. Now try it on Ax=b where

```
A=triu(tril(rand(10),1),-1)
b=(1:10)'.
```

10. Each time you apply an elerow3 it is the same as multiplying by an ele3 (see page 9.) Thus in your trisolve you are getting

$$E_{n-1} \cdots E_2 E_1 A = U$$

where each E_i is an ele3 and U is upper triangular. Let $L = (E_{n-1} \cdots E_2 E_1)^{-1}$, then $A = LU$. Now each of the E_i is lower triangular, the product of these is also lower triangular, and so is the inverse L. For an explanation of these statements about lower triangular matrices see *The LU Decomposition*. Using this formula write a MATLAB function [L,U]=trilu(A) which finds an upper triangular U and a lower triangular L where A=L*U.

11. Check your trilu on

A=tril(triu(rand(8),-1),1) and

A=tril(triu(ones(8),-1),1)

To see if you are on the right track, look at A-L*U.

COMMENTS

While MATLAB's A\b is an excellent general purpose solver, special classes of matrices like the triangular matrices and the tridiagonal matrices permit faster methods. To see how we measure speed see *Flops*.

The LU Decomposition

ABSTRACT

In the process of row reducing a matrix it is possible to get a decomposition of the matrix into a product of an upper triangular and a lower triangular matrix. This is a useful decomposition for solving systems. This project assumes that you have created the file elerow3.m (see page 30 of *Systems of Linear Equations.*)

MATLAB COMMANDS

    lu, tril, triu, diag, elerow3, *, ', \

LINEAR ALGEBRA CONCEPTS

Gaussian Elimination, Triangular Matrix, Elementary Matrix, Permutation Matrix, Positive Definite Matrix

BACKGROUND

Suppose that A is an $n \times n$ matrix and that after applying a sequence of elementary row operations of type 3 to A we have arrived at an upper triangular matrix U. This process, called Gaussian Elimination, is given by the following simple program, badgauss. The purpose of badgauss is to illustrate the structure of Gaussian Elimination, since it does no row switches it may try to divide by 0.

```
function A=badgauss(A)

% A=badgauss(A). Given a square
% matrix A badgauss returns a
% matrix in triangular form. It
% does no row switching.

[m,n]=size(A);
for j=1:n-1,
 for i=j+1:n,
  A=elerow3(A,-A(i,j)/A(j,j),j,i);
 end
end
```

Since each elementary row operation corresponds to matrix multiplication by an elementary matrix, we have

$$E_k \ldots E_1 A = U$$

where each E_i is an elementary matrix of type 3. Since the elementary matrices are invertible we can solve for A by multiplying by the reverse sequence of matrix inverses. Thus

$$A = E_1^{-1} \ldots E_k^{-1} U$$

with the inverse matrices being determined by the rule: if $E = \text{ele3}(n, r, i, j)$, then $E^{-1} = \text{ele3}(n, -r, i, j)$. See *Building Matrices*, page 9, for the notation ele3. We have the result

Theorem 1. *LU Decomposition. If A reduces to an upper triangular matrix U by elementary row operations of type 3, then there is a lower triangular matrix L such that*

$$A = LU.$$

PROOF: We only need to show that $E_1^{-1} \ldots E_k^{-1}$ is a lower triangular matrix. We have seen above each of these is an elementary matrix of type 3 of the form $\text{ele3}(n, r, i, j)$, but note further that from the algorithm we know that $i < j$. Thus each of these elementary matrices is lower triangular. The proof can be finished by showing that the product of lower triangular matrices is lower triangular and thus we can take $L = E_1^{-1} \ldots E_k^{-1}$. ∎

Given $A = LU$ we can solve the system $Ax = b$ by first solving $Ly = b$ using forward substitution and solving $Ux = y$ by back substitution. See *Triangular Systems* for back substitution.

$$Ax = LUx = Ly = b$$

In Gaussian Elimination if $A(j, j) = 0$ it is necessary to interchange rows using an elementary row operation of type 1 in order to move a nonzero entry into the (j, j) position (look at line 4 of badgauss to see that otherwise you would be dividing by 0.) This row switch is the same as multiplying the matrix by an $\text{ele1}(n, i, j)$ for some i.

It turns out that it is necessary to perform row interchanges for numerical reasons as well. For a discussion of the stability of algorithms see *Errors*. Consider the following example of Forsythe and Moler. In MATLAB type

```
A=[.1*eps, 1; 1, 1]; b=[1; 2];
C=[A,b];
C=elerow3(C,-C(2,1)/C(1,1),1,2);
```

Now we will do back substitution to find the solution

```
y=C(2,3)/C(2,2);
x=(C(1,3)-C(1,2)*y)/C(1,1);
```

But notice that $x \approx 1, y \approx 1$ is a much more accurate solution than this one. This problem disappears if we perform a row interchange first and then apply the elementary row operation of type 3. Rebuild C as above but do the following row switch before the type 3 elementary row operation.

```
A=[.1*eps, 1; 1, 1]; b=[1; 2];
C=[A,b];
C([1,2],:)=C([2,1],:);
C=elerow3(C,-C(2,1)/C(1,1),1,2);
y=C(2,3)/C(2,2);
x=(C(1,3)-C(1,2)*y)/C(1,1);
```

This is a much more satisfying answer. Notice that MATLAB's $A \backslash b$ gives the same answer.

Pivoting Strategies

An approach to correcting this type of error is the ***partial pivoting strategy***. See page 26 for the definition of ***pivot***. On the jth iteration of the Gaussian Elimination instead of automatically pivoting on $A(j, j)$, choose $k \geq j$ such that

$$|A(k, j)| = \max_{i \geq j} |A(i, j)|$$

and pivot on $A(k, j)$ instead. We are faced with the prospect of row interchanges, *i.e.* multiplication by elementary matrices of type 1.

Unfortunately, partial pivoting has not solved all of our problems. In the previous example rebuild C.

```
A=[.1*eps, 1; 1, 1]; b=[1; 2];
C=[A,b];
```

Now multiply the first equation by

```
1/C(1,1);
C(1,:)=(1/C(1,1))*C(1,:)
```

Now partial pivoting will allow us to pivot on the (1,1) entry to get

```
C=elerow3(C,-C(2,1)/C(1,1),1,2);
```

and using back substitution

```
y=C(2,3)/C(2,2);
x=(C(1,3)-C(1,2)*y)/C(1,1);
```

As you can see we have fallen into the same trap. Here the matrix is badly ***scaled*** and this can be (at least partially) corrected with a technique called ***scaled partial pivoting***. We compute the scaling factors for each row

$$s_i = \max_{1 \leq j \leq n} |A(i, j)|$$

before finding the pivot. When it is time to find the pivot element choose $k \geq i$ so that

$$|A(k, i)/s_i| = \max_{j \geq i} |A(j, i)/s_j|.$$

In the case of our example, the scaled partial pivoting technique tells us to pivot on the (2, 1) entry. After rebuilding C we get

```
C([1,2],:)=C([2,1],:);
C=elerow3(C,-C(2,1)/C(1,1),1,2);
```

Now you can see that we will get a reasonable answer. Try MATLAB's \ on this example.

The pivoting strategies introduce elementary row operations of type 1 and hence, ele1's, into the factorization. This changes the nature of the decomposition a bit. A product of ele1's is called a ***permutation matrix***. As you might expect, if P is a permutation matrix, then PA rearranges the rows of A. It is easy to compute the inverse of a permutation matrix, if $P = E_1 \ldots E_k$ then

$$P^{-1} = E_k^{-1} \ldots E_1^{-1} = E_k^T \ldots E_1^T = (E_1 \ldots E_k)^T = P^T$$

since $E_i^{-1} = E_i^T$ for elementary matrices of type 1. The appropriate decomposition is the PLU Decomposition. Since a proof of this is rarely seen, we include it here.

Theorem 2. *The PLU Decomposition. If A is any $n \times n$ matrix, then there exist an upper triangular matrix U, a lower triangular matrix L, and a permutation matrix P, such that*

$$PA = LU.$$

PROOF: The proof which we will give is by induction on the size of the matrix n. If $n = 1$ there is nothing to prove since A is upper triangular so we can take $U = A$, $L = I_n$ and $P = I_n$. Though it is not necessary we will do the $n = 2$ case to get some feeling for the method. If $A(1, 1) \neq 0$ then we can use Gaussian Elimination without row interchanges just as in Theorem 1, and again take $P = I_2$. If $A(1, 1) = 0$ and $A(1, 2) \neq 0$ then perform a row interchange by multiplying by an elementary matrix of type 2, E and apply Theorem 1 to the matrix EA to get $EA = LU$ and now we take $P = E$.

Now suppose by induction that we know the PLU Theorem works for matrices of size $n - 1$ and suppose that A has size n. Use an elementary matrix E of type 2 to get a nonzero entry in the $(1, 1)$ position of EA. Write

$$EA = \begin{bmatrix} a & v \\ u & A_1 \end{bmatrix}$$

where $a \neq 0$, v is $1 \times (n - 1)$, u is $(n - 1) \times 1$, and A_1 is $(n - 1) \times (n - 1)$. The matrix $-uv/a + A_1$ is $(n-1) \times (n-1)$ (uv is an *outer* product, see page 32) so by induction, apply the PLU Decomposition to $-uv/a + A_1$ to get a permutation matrix P_1, a lower triangular matrix L_1 and an upper triangular matrix U_1 with

$$-uv/a + A_1 = P_1^T L_1 U_1$$

Now notice that

$$EA = \begin{bmatrix} a & v \\ u & A_1 \end{bmatrix} = \begin{bmatrix} 1 & \vec{0} \\ \vec{0} & P_1^T \end{bmatrix} \begin{bmatrix} 1 & \vec{0} \\ P_1(u/a) & L_1 \end{bmatrix} \begin{bmatrix} a & v \\ \vec{0} & U_1 \end{bmatrix}$$

Thus we can take

$$P = \begin{bmatrix} 1 & \vec{0} \\ \vec{0} & P_1 \end{bmatrix} E, \quad L = \begin{bmatrix} 1 & \vec{0} \\ P_1(u/a) & L_1 \end{bmatrix}, \quad U = \begin{bmatrix} a & v \\ \vec{0} & U_1 \end{bmatrix}$$

∎

The proof of this theorem lends itself to a recursive program but it is not a recommended method of programming the decomposition (see Problem 4). MATLAB has a function which computes the LU Decompostion.

```
A=ones(5)+diag(1:5)
[L,U]=lu(A),
A-L*U
```

The lower triangular matrix L obtained always has 1's on the main diagonal. This permits a storage method which holds U on and above the diagonal and L below the diagonal. The call B=lu(A) will return a matrix B which stores L and U in this manner. The call [L,U]=lu(A) will return two matrices; with U an upper triangular matrix though the L will not necessarily be lower triangular – the rows need to be permuted. But it is the case that A=L*U.

```
B=lu(A)
triu(B)-U
(-tril(B,-1)+eye(5))- L
```

But notice that L may not be lower triangular

```
A=magic(5);
[L,U]=lu(A)
```

By permuting the rows of L we can make it into a lower triangular matrix. In this example we see that by switching rows 1 and 2, 3 and 5, and 4 and 5, in that order, we will get a lower triangular matrix. This is done with a permutation matrix

```
P=eye(5);
P([1,2],:)=P([2,1],:)
P([3,4,5],:)=P([5,3,4],:)
```

You can see that this has the desired effect

```
P*L
```

The PLU Decomposition is now obtained by replacing L with P*L

```
L=P*L; P*A-L*U
```

We solve a system of equations using the PLU Decomposition in basically the same way as with an LU Decompositon. Suppose that $PA = LU$ and $Ax = b$ is to be solved, then $A = P^{-1}LU$. Now solve $Ly = Pb$ and $Ux = y$, then

$$Ax = P^{-1}LUx = P^{-1}Ly = P^{-1}Pb = b.$$

The Cholesky Factorization

Recall that A is **symmetric** if $A = A^T$. A is **positive definite** if A is symmetric and $x^T Ax > 0$ for all $x \neq \vec{0}$. An application of positive definite matrices is given in *Symmetric Diagonalization*. An easy way to produce symmetric positive definite matrices is to start with a nonsingular B and let $A = BB^T$. Notice that $A^T = (BB^T)^T = B^{TT}B^T = BB^T = A$, so that A is symmetric. If we let x be any vector, $x^T Ax = x^T BB^T x = Bx \cdot Bx \geq 0$. The dot product is > 0 provided $Bx \neq \vec{0}$. Since B is invertible, $Bx \neq \vec{0}$ when $x \neq \vec{0}$.

Theorem 3. *Suppose that A is a symmetric matrix which can be reduced to triangular form using only elementary row operations of type 3 and Gaussian Elimination. Then there is a lower triangular matrix L and a diagonal matrix D such that $LAL^T = D$.*

PROOF: Suppose that $E_k \ldots E_1 A = U$ where each E_i is a lower triangular elementary matrix of type 3. Form

$$D = E_k \ldots E_1 A E_1^T \ldots E_k^T.$$

This performs the corresponding column operations on A. Notice that since A is symmetric, so is D. The operations $E_1^T \ldots E_k^T$ all work on the upper triangle and so the operations $E_k \ldots E_1$ still bring A into upper triangular form. Since D is symmetric and triangular, it is a diagonal matrix. Let $L = E_k \ldots E_1$. ∎

In view of Theorem 3 our goal is to show that for a positive definite symmetric matrix we can reduce to a diagonal D which has positive diagonal entries using only type 3 operations.

Theorem 4. *Cholesky Factorization. If A is symmetric positive definite, then there is a lower triangular matrix G with positive entries on the diagonal such that $A = GG^T$.*

PROOF: Let $e1$ be given by

$$e1 = \begin{bmatrix} 1 \\ 0 \\ \vdots \\ 0 \end{bmatrix}$$

Since A is positive definite, $e_1^T A e_1 = a_{11} > 0$, thus we can use a_{11} as a pivot. By the technique of Theorem 3, there is a lower triangular matrix L_1 such that

$$L_1 A L_1^T = \begin{bmatrix} a_{11} & 0 & \cdots & 0 \\ 0 & & & \\ \vdots & & B & \\ 0 & & & \end{bmatrix}$$

Now B is also symmetric positive definite and so the same idea can be applied repeatedly. This gives a lower triangular L with 1's on the diagonal such that $LAL^T = D$ is diagonal and you will notice that the diagonal entries are all positive. Let $D^{1/2}$ be the result of taking the square roots of the diagonal entries of D, and let $G = L^{-1}D^{1/2}$, then $GG^T = L^{-1}D^{1/2}(L^{-1}D^{1/2})^T = L^{-1}D(L^T)^{-1} = A$. G is lower triangular and the diagonal entries are positive. ∎

The Cholesky Factorization is unique, that is, if A is positive definite, and $A = GG^T = LL^T$ where L and G are lower triangular, then $G = L$. MATLAB has a built-in function chol which finds the Cholesky factorization. A call to U=chol(A) returns an upper triangular matrix U where A=U'*U.

```
A=rand(5);  A=A'*A;
U=chol(A)
A-U'*U
```

PROBLEMS

1. Using MATLAB's lu find the PLU decomposition of A=magic(7). You will need to find a lower triangular matrix L (since lu will not return a lower triangular matrix) and a permutation matrix P so that P*A=L*U.

2. Let A=magic(7) and let b=A*(1:7)'. Using the PLU decomposition from problem 1 solve the system Ax=b using forward substitution and backward substitution. Since you know the correct answer you can check your work.

3. A permutation matrix is an example of a sparse matrix, that is, a matrix with few nonzero entries. We usually try to find alternative ways of storing large sparse matrices to save memory. An $n \times n$ permutation matrix can be stored as a $1 \times n$ vector. Find a vector *piv* which codes the permutation matrix P in problem 1 in the following sense: if we let $c = b(piv)$, then we can solve $Ax = b$ by solving $LUx = c$. (You may want to review the *MATLAB Tutorial* to recall how $c = b(piv)$ works. See the subsection on Colon Notation.)

4. Write a MATLAB function [P,L,U]=plu(A) which returns P, L, U with the property that P*A=L*U using recursion which is based on the proof of Theorem 2. Test your plu on the example in problem 1. Is the PLU Decomposition unique? Modify your plu to do scaled partial pivoting. Does the modified version return the same values as the original in the example? How does it compare with MATLAB's lu?

5. A matrix is ***diagonally dominant*** if $|A(i, i)| > \sum_{j \neq i} |A(i, j)|$. The importance of a diagonally dominant matrix here is that it is not necessary to use any pivoting strategies, so that badgauss should work for diagonally dominant matrices. Type in badgauss and try it on

```
A=rand(5)+100*diag(rand(5,1))
```

How does badgauss compare with lu on this example?

6. You can create a symmetric positive definite matrix with known Cholesky factorizationby starting with an upper triangular matrix with positive entries on the diagonal, say U=triu(rand(7)) and letting A=U'*U. Find the Cholesky Factorization, R=chol(A). Since the Cholesky Factorization is unique, try R-U.

7. Write a MATLAB function b=spd(A) which returns b=1 if A is symmetric positive definite and b=0 otherwise. Do not use chol. You can use badgauss to get started since we know from the proof of Theorem 4 that we do not need to do any pivoting for a symmetric positive definite matrix. Check your spd on A=ones(5)+diag(1:5) and the matrix in problem 5. You can use chol to check your answers.

8. The normal equations (see page 125), that is, a matrix equation of the form $A^T Ax = A^T b$ are important in least squares problems. See *Least Squares* for a discussion about least squares problems. Getting accurate solutions rapidly is important to least squares problems. When rank$(A) = n$, where n is the number of columns of A, $A^T A$ is a symmetric positive definite matrix. Let A=magic(8); A=A(:,1:3). Check that rank(A)=3. Now apply a Cholesky factorization to A'*A to get A'*A=R'*R. Solve A'*Ax=R'*Rx=A'b where R=chol(A'*A) and b=ones(8,1) using forward and back substitution.

Inverses

ABSTRACT
The inverse of a matrix is defined and methods are proposed for computing inverses.

MATLAB COMMANDS
```
inv, norm, *, +, -, \
```

LINEAR ALGEBRA CONCEPTS
Triangular Matrix, Back Substitution, Elementary Matrix, Norm

BACKGROUND

Given an $n \times n$ matrix A, the **inverse** of A is the matrix B such that $AB = BA = I_n$. It is not always the case that a matrix has an inverse. When A has an inverse we say A is **invertible** or **nonsingular** and write the unique inverse as A^{-1}. If A has no inverse, we say it is **singular** or **noninvertible.** MATLAB has a function inv(A) which computes the inverse of a nonsingular matrix.

The utility of an inverse comes from the fact that we can get an explicit form of the unique solution to $Ax = b$ if A is invertible, since $x = A^{-1}b$. This is useful when it is necessary to solve several equations $Ax = b_1, \ldots, Ax = b_k$. Under these conditions it may be worthwhile to have the inverse to solve each of these equations. In general, computing the inverse takes much more time than simply solving the equation directly. While x=inv(A)*b does give the solution to Ax=b, a faster way is to use the MATLAB solver $A \backslash b$.

The inverse of AB can be computed from inverses for A and B. Namely, $(AB)^{-1} = B^{-1}A^{-1}$, since $(AB)(B^{-1}A^{-1}) = AI_nA^{-1} = I_n$.

The most direct approach to finding the inverse is to write $A^{-1} = [x_1, x_2, \ldots, x_n]$ where $x_1, \ldots, x_n$ are column vectors which are to be determined. Now

$$AA^{-1} = [Ax_1, \ldots, Ax_n] = I_n = [e_1, \ldots, e_n]$$

where e_i is the standard basis vector (see page 9 of *Basics: Vectors and Matrices*.) We can find x_i by solving $Ax_i = e_i$. This is easily done in MATLAB using $x_i = A \backslash e_i$. This is not the best way, since MATLAB recomputes the Gaussian Elimination for each equation $Ax_i = e_i$.

There is an explicit formula for the inverse in terms of the determinant but it is inefficient to compute and it is primarily of theoretical interest.

Inverses and Row Reduction

Invertibility is closely related to the elementary row operations. Recall from page 33 of *Systems of Linear Equations* that

(1) $\text{ele1}(n, i, j)^{-1} = \text{ele1}(n, i, j)$
(2) $\text{ele2}(n, r, i)^{-1} = \text{ele2}(n, 1/r, i)$
(3) $\text{ele3}(n, r, i, j)^{-1} = \text{ele3}(n, -r, i, j)$

Theorem 1. *A is invertible if and only if A row reduces to I_n.*

Assuming that A is invertible, then we can row reduce A to I_n, and so there are elementary matrices $E_1, \cdots E_k$, such that $E_1 \cdots E_k A = I_n$. Thus $A^{-1} = E_1 \cdots E_k$. A way to find A^{-1} is to reduce $[A, I_n]$ since

$$E_1 \cdots E_k[A, I_n] = [I_n, E_1 \cdots E_k] = [I_n, A^{-1}]$$

This method works well, though it takes additional time to bring $[A, I_n]$ into reduced row echelon form. It also uses additional space to store the augmented matrix, which is wasteful when working with very large matrices.

A Newton Style Iteration

An iterative method for finding the inverse is given by an adaptation of Newton's Method. Let B_0 be an initial guess at the inverse and now define the sequence $B_0, B_1, B_2, \ldots$ by

$$B_{m+1} = B_m + B_m(I_n - AB_m)$$

Define the residual $R_m = I_n - AB_m$, to get the formula $B_{m+1} = B_m + B_m R_m$. Notice that

$$
\begin{aligned}
R_{m+1} &= I_n - AB_{m+1} \\
&= I_n - AB_m + AB_m R_m \\
&= R_m - AB_m R_m \\
&= R_m(I_n - AB_m) \\
&= R_m^2
\end{aligned}
$$

and thus $R_m = R_0^{2^m}$. The next theorem provides the conditions which permit this method to converge. The notation $\|A\|$ refers to the norm of A which is computed in MATLAB with `norm`.

Theorem 2. *If $\|R_0\| < 1$, then $\lim_{m \to \infty} B_m = A^{-1}$*

PROOF: The norm $\|A\|$ (see page 83 of *Norms and Condition Numbers*) is a number assigned to a matrix with these properties:

$\|AB\| \le \|A\| \|B\|$

$\|A\| = 0$ implies $A = 0$.

If $\|R_0\| < 1$, then since $\|R_m\| = \|R_0\|^{2^m}$, we get $\lim_{m \to \infty} R_m = 0$. Let $B = \lim_{m \to \infty} B_m$, then

$$
\begin{aligned}
I_n - AB &= \lim_{m \to \infty} I_n - AB_m \\
&= I_n - A(\lim_{m \to \infty} B_m) \\
&= \lim_{m \to \infty} R_m = 0.
\end{aligned}
$$

So $I_n = AB$ and $B = A^{-1}$. ∎

The tricky part is to choose B_0 so that $\|R_0\| < 1$. Try this in MATLAB

```
a=magic(5)
b=inv(a)+.001*rand(5)
```

Check that this b is a good B_0

```
r=eye(5)-a*b
norm(r)
```

Set up the iteration

```
b=b+b*(eye(5)-a*b)
```

Now, see how close this is to the inverse

```
b-inv(a)
```

Repeat the following several times

```
b=b+b*(eye(5)-a*b), b-inv(a)
```

You should see b converge. Try

```
a*b
```

The LU Decomposition

The MATLAB function `inv(A)` computes the inverse of a nonsingular matrix using the LU Decomposition of a matrix. The LU Decomposition is a factorization of $A = LU$ where L is lower triangular (see page 90) with 1's on the diagonal and U is upper triangular. Usually, the L is not really lower triangular, but a lower triangular matrix with its rows switched around. See *The LU Decomposition* for the details of the LU Decomposition. The algorithm to perform this decomposition is a highly accurate form of Gaussian Elimination. Now suppose that $A = LU$, then $A^{-1} = U^{-1}L^{-1}$ so that if we let $X = U^{-1}L^{-1}$ we see that $UX = L^{-1}$ and $XL = U^{-1}$. We will show how to use this to solve for X when A is 3×3. From $XL = U^{-1}$ and the fact that U^{-1} is upper triangular, we get the following three equations

$$X(2, :) * L(:, 1) = 0 \quad X(3, :) * L(:, 1) = 0 \quad X(3, :) * L(:, 2) = 0$$

From $UX = L^{-1}$ and the fact that L^{-1} is lower triangular we get the following six equations

$$U(1, :) * X(:, 1) = 1 \quad U(1, :) * X(:, 2) = 0 \quad U(1, :) * X(:, 3) = 0$$

$$U(2, :) * X(:, 2) = 1 \quad U(2, :) * X(:, 3) = 0 \quad U(3, :) * X(:, 3) = 1$$

If you write these equations out, you will discover that they can be arranged to form an upper triangular system, which can be solved using back substitution (see page 89.)

PROBLEMS

1. Try MATLAB's `inv(A)` on the following matrices: `rand(5)`, `hilb(8)`, `magic(9)`, `pascal(9)`. Look at the residual `eye(n)-A*inv(A)` in each case. In general the residual is a poor check for accuracy. `invhilb(n)` is a function which finds only the inverse of the Hilbert matrix `hilb(n)`. Compare the result of `inv(hilb(10))` and `invhilb(10)` and look at the residual of each.

2. Find the inverse of `A=magic(5)` by solving the equations `Ax=E(:,1),...,` `Ax=E(:,5)` where `E=eye(5)` and glueing the results into `B=[x1,...,x5]`. How does this compare with `inv(A)`?

3. Use the method of reducing $[A, I_n]$ to find the inverse of A of the matrices in problem 1. Check the method for accuracy against `inv`. A quick way to check how close a matrix `C` is to a matrix `D` is `max(max(abs(C-D)))`.

4. Try a Newton-style iteration on

$$A=4*\text{eye}(5)+\text{diag(ones}(4,1),1)+\text{diag(ones}(4,1),-1)$$

Choose `B=.25*eye(5)`. Why is this a reasonable choice? Compute the norm on each iteration with `norm(eye(5)-A*B)`. To get some idea of how sensitive this method is to the choice of B, let `A=magic(5)`. First let `B=inv(A)+.001`. Now try `B=inv(A)+.01`.

5. Use the LU Decomposition to find the inverse by back substitution. Let `A=LU` where `L=eye(3)+tril(magic(3),-1)` and `U=triu(rand(3))`. First find the nine equations and arrange them in an upper triangular system. Then use `backsub` (see page 91) to find the solution. Finally, arrange the solution properly in the matrix. How does this compare with `inv(A)`?

6. Write a MATLAB function `B=upinv(A)` which computes the inverse of an upper triangular matrix by solving the equations `Ax=E(:,i)` where `E=eye(n)` for `i=1:n` using back substitution.

7. Try your `upinv` on `triu(rand(10))`. Try it on the Wilkinson matrix in problem 3 of *Triangular Systems*. How does it compare with `inv`? Look at the residuals in each case. Which one is the more accurate inverse?

8. For those who are familiar with MATLAB's `flops` (see *Flops*) obtain flop counts using the Newton Style Iteration for `magic(5)` in and using `inv`. Which is faster?

9. When the series $I + A + A^2 + A^3 + \cdots$ converges, it converges to the inverse of I-A, that is,

$$(I - A)^{-1} = I + A + A^2 + A^3 + \cdots$$

You can see this happen in MATLAB. Let

```
A=rand(5)/5; B=eye(5)-A; C=inv(B);
```

We are going to find the inverse of B. We can compare the series to C.

```
INV=eye(5);
     for i=1:20, INV=eye(5)+A*INV; norm(INV-C,inf), end
```

This does not work for all matrices, since the series will not converge for all choices of A. Try this on `A=rand(5)`.

Subspaces

ABSTRACT

The concepts of linear combination, subspace, column space, and row space are introduced.

MATLAB COMMANDS

```
rref, *
```

LINEAR ALGEBRA CONCEPTS

Linear Combination, Subspace, Span, Column Space, Row Space

BACKGROUND

Suppose that $v_1, \ldots, v_n$ are vectors of length m, then a ***linear combination*** of $v_1, \ldots, v_n$ is a sum

$$r_1 v_1 + r_2 v_2 + \ldots + r_m v_m$$

where $r_1, \ldots, r_m$ are scalars. While this looks new, the next theorem shows that a linear combination is really just the outcome of a vector multiplied by a matrix. This theorem is crucial in using MATLAB to solve problems in the algebra of vectors.

Theorem 1. *Suppose that $A = [v_1, \ldots, v_n]$ where $v_1, \ldots, v_n$ are the columns of A, then*

$$Ar = r_1 v_1 + r_2 v_2 + \cdots + r_n v_n$$

where $r = [r_1, \ldots, r_n]^T$.

PROOF: Suppose that the (i, j) entry of A is a_{ij}. Then

$$
\begin{aligned}
Ar &= \begin{bmatrix} a_{11} r_1 + \ldots + a_{1n} r_n \\ \vdots \\ a_{m1} r_1 + \ldots + a_{mn} r_n \end{bmatrix} \\
&= r_1 \begin{bmatrix} a_{11} \\ \vdots \\ a_{m1} \end{bmatrix} + \ldots + r_n \begin{bmatrix} a_{1n} \\ \vdots \\ a_{mn} \end{bmatrix} \\
&= r_1 v_1 + r_2 v_2 + \ldots + r_m v_m
\end{aligned}
$$

∎

Try this in MATLAB

```
A=hilb(4)
r=(1:4)'
A*r
r(1)*A(:,1)+r(2)*A(:,2)+r(3)*A(:,3)+r(4)*A(:,4)
```

105

Notice that $A\vec{0} = \vec{0}$ always holds, so $\vec{0}$ is always a linear combination of the columns of A. $W \subseteq R^m$ is a **subspace** of R^m if $W \neq \emptyset$ and for all vectors $v_1, \ldots, v_k \in W$ and all scalars $r_1, \ldots, r_k$ the linear combination

$$r_1 v_1 + \ldots + r_k v_k \in W.$$

The customary definition of a subspace is given by the three properties
   (1) $\vec{0} \in W$
   (2) if $u, v \in W$, then $u + v \in W$
   (3) if $u \in W$ and r is any scalar, then $ru \in W$.
The two definitions can be shown to be equivalent.

The easiest examples of subspaces are $W = R^n$ and $W = \{\vec{0}\}$. Another way to generate examples of subspaces is to look at the set of all linear combinations of some vectors $v_1, \ldots, v_n$. Define the **span of** $v_1, \ldots, v_n$ by

$$\text{span}(v_1, \ldots, v_n) = \{r_1 v_1 + \ldots + r_n v_n : r_1, \ldots r_n \text{ are scalars}\}.$$

$\text{span}(v_1, \ldots, v_n)$ is the **the subspace generated by** $v_1, \ldots, v_n$.

Theorem 2. *If $v_1, \ldots, v_n \in R^m$, then*
(1) $W = \text{span}(v_1, \ldots, v_n)$ is a subspace of R^m.
(2) If U is any subspace with $v_1, \ldots, v_n \in U$, then $\text{span}(v_1, \ldots, v_n) \subseteq U$.

The Column Space and the Row Space

If $A = [v_1, \ldots, v_n]$, then $\text{span}(v_1, \ldots, v_n)$ is called the **Column Space of A**. The **Row Space of A** is $\text{span}(w_1, \ldots, w_m)$, where

$$A = \begin{bmatrix} w_1 \\ \vdots \\ w_m \end{bmatrix}$$

and $w_1, \ldots, w_m$ are the rows of A. A modification of Theorem 1 works for the Row Space.

Theorem 3. *Suppose that $w_1, \ldots, w_m$ are the rows of A, then*

$$rA = r_1 w_1 + r_2 w_2 + \ldots + r_m w_m$$

where $r = [r_1, \ldots, r_m]$.

Note: r is a $1 \times m$ row vector being multiplied on the left hand side of A.

How do we test if a column vector u is in the Column Space of A? We try to solve

$$u = r_1 v_1 + \ldots + r_n v_n$$

for $r_1, \ldots, r_n$ where $v_1, \ldots, v_n$ are the columns of A. By Theorem 1, $u = r_1v_1 + \ldots + r_nv_n = Ar$. For a matrix which is either singular or not square, I suggest that you use rref. In MATLAB try

```
A=magic(4);
b=ones(4,1)
rref([A,b])
```

So we see that Ax=b has a solution, and thus b is in the Column Space of A. Now let

```
b=(1:4)'
rref([A,b])
```

Now we see that Ax=b does not have a solution, and thus b is not in the Column Space of A.

How do we test if a row vector u is in the Row Space of A? Again we try to solve

$$u = r_1w_1 + r_2w_2 + \ldots + r_mw_m$$

where $w_1, \ldots, w_m$ are the rows of A. By Theorem 3, this amounts to solving $rA = u$, or we can take transposes, and solve $A^T r^T = u^T$.

Theorem 4. *If w is a linear combination of $v_1, \ldots, v_n$, then*

$$span(v_1, \ldots, v_n, w) = span(v_1, \ldots, v_n).$$

Theorem 4 suggests that we can systematically remove some of the vectors in the list $v_1, \ldots, v_n$ to achieve a "minimal spanning set of vectors." This is true and one way to go about it is to ask first if v_1 is a linear combination of $v_2 \ldots, v_n$, if so remove it, then go on to ask if v_2 is a linear combination of the remaining vectors, etc. There is a better way. Suppose that A row reduces to R which is in reduced row echelon form. By looking at R, we see that any column which does not contain a leading 1 is a linear combination of the columns which do contain leading ones. For example, suppose that $A = [v_1, v_2, v_3, v_4]$ and A reduces to

$$\begin{bmatrix} 1 & 2 & 0 & 3 \\ 0 & 0 & 1 & 4 \\ 0 & 0 & 0 & 0 \end{bmatrix}$$

We get nontrivial solutions to $\alpha_1v_1 + \alpha_2v_2 + \alpha_3v_3 + \alpha_4v_4 = \vec{0}$ by letting $\alpha_2 = 1$ and $\alpha_4 = 0$ to get $-2v_1 + v_2 = \vec{0}$, so v_2 is a linear combination of v_1. Letting $\alpha_2 = 0$ and $\alpha_4 = 1$ we get $-3v_1 - 4v_3 + v_4 = \vec{0}$. So v_4 is a linear combination of v_1 and v_3. Thus we can remove v_2 and v_4 and for the remaining vectors v_1 and v_3, $span(v_1, v_3) = span(v_1, v_2, v_3, v_4)$. If B is the matrix consisting of the columns of A which become the columns of leading 1's in R, then the Column Space of A = Column Space of B. Can any more columns be removed? Why? We can handle this easily in MATLAB using the function lead (see page 27 of *Systems of Linear Equations.*)

```
B=rref(A)
```

```
L=lead(B)
B=A(:,L)
```

From above we know that the Column Space of A = Column Space of B. There is another approach which applies to the Row Space.

Theorem 5. *If B is obtained from A by applying elementary row operations, then the Row Space of B = Row Space of A. It follows that if B consists of the nonzero rows of the reduced row echelon form, then the Row Space of B = Row Space of A*

Note: This theorem is not true for column spaces. By taking transposes, it is possible to use this theorem on a column space, since a vector in the Column Space of A is just the transpose of a vector in the Row Space of A^T. Theorem 5 provides us with a way to get a "minimal spanning set" for a row space, simply by removing the rows of zeros from `rref(A)`.

If we have two subspaces V and W of R^n we can combine them to get new subspaces. The intersection, $V \cap W$, and the sum, $V + W$ are defined by

$$V \cap W = \{u \in \mathrm{R}^n : u \in V \text{ and } u \in W\}$$

and

$$V + W = \{v + w : v \in V \text{ and } w \in W\}$$

Theorem 6. *If V and W are subspaces of R^n, then $V \cap W$ and $V + W$ are subspaces of R^n.*

Now if V and W are the Column Spaces of A and B, respectively, then for $v \in V$ and $w \in W$, v is a linear combination of the columns of A and w is a linear combination of the columns of B and so $v + w$ is in the Column Space of $[A, B]$. It follows that the Column Space of $[A, B]$ is $V + W$. Similarly, if V and W are the Row Spaces of A and B, respectively, then $V + W$ is the Row Space of $[A; B]$. Handling the intersection of column spaces is more complicated and we will defer this until *The Null Space*.

PROBLEMS

1. Determine if u=(1:5)' is in the Column Space of A for each of the following matrices: `magic(5)`, `list(5)`, `rand(5)` and A=v*w where v=rand(5,1) and w=rand(1,5). See page 34 of *Building Matrices* for `list`.

2. Determine if u=(1:5) is in the Row Space of A for each of the matrices in problem 1.

3. For each of the matrices in problem 1, find a submatrix B of the matrix A where the Column Space of A = Column Space of B and no column of B is a linear combination of the other columns of B. You may use the function `lead` from page 27 of *Systems of Linear Equations*.

4. Write a MATLAB function B=shrink(A) which returns a submatrix B of A with the Column Space of A = Column Space of B and no column of B is a linear combination of the other columns of B. Compare your shrink with the answers from problem 3.

5. For each of the matrices in problem 1, find a matrix B where the Row Space of A = the Row Space of B and no row of B is a linear combination of the other rows of B. Use Theorem 5 to solve this problem.

6. Use the function shrink to do problem 5 by using the fact that a vector in the Row Space of A is the transpose of a vector in the Column Space of A. You may not get the same vectors, but you should get the same number of vectors.

7. Write a MATLAB function B=rowshrk(A) which returns a matrix B with the Row Space of B= Row Space of A and no row of B is a linear combination of other rows of B. Compare your rowshrk with the answers from problem 5.

8. Show, by finding an example, that if A row reduces to R, then the Column Space of A need not be the Column Space of R. Hint: Some of the matrices in problem 1 will work here.

9. Suppose that the columns of A are a minimal spanning set for V and the columns of B are a minimal spanning set for W. Show, by finding an example, that the columns of $[A, B]$ need not be a minimal spanning set for $V + W$. Hint: Some of the matrices in problem 1 will work here.

Dimension and Rank

ABSTRACT
Introducing the concepts of linear independence, basis, dimension, and rank.

MATLAB COMMANDS
```
rref, rank
```

LINEAR ALGEBRA CONCEPTS
Linear Independence, Rank, Basis, Dimension, Row Space, Column Space, Coordinates

BACKGROUND
Suppose that $v_1, \ldots, v_n \in \mathbb{R}^m$. We say that $v_1, \ldots, v_n$ are **linearly independent** if

$$r_1 v_1 + \ldots + r_n v_n = \vec{0} \quad \text{implies} \quad r_1 = \ldots = r_n = 0.$$

This can be rephrased as follows: let $A = [v_1, \ldots, v_n]$ and $r = [r_1, \ldots, r_n]^T$, then

$$Ar = r_1 v_1 + \ldots + r_n v_n$$

(See Theorem 1 in *Subspaces*.) Thus $v_1, \ldots, v_n$ are linearly independent when

$$Ar = \vec{0} \text{ implies } r = \vec{0}.$$

It follows that $v_1, \ldots, v_n$ are linearly independent if and only if $Ax = \vec{0}$ has a unique solution, that is,

$$A \text{ reduces to } \begin{bmatrix} I_n \\ 0 \end{bmatrix}$$

We see that $\vec{0}$ is never in a set of linearly independent vectors. The easiest example of linearly independent vectors is $e_1, \ldots, e_m$. Here is another example from MATLAB.

```
A=rand(6,4);
rref(A)
```

Rank

The **rank of** A is defined to be the number of nonzero rows in `rref(A)`. MATLAB computes this with `rank`. We can now say that the columns of an $m \times n$ matrix A are linearly independent if and only if $\text{rank}(A) = n$. Notice also that

$$Ax = b \text{ has a solution if and only if } \text{rank}(A) = \text{rank}([A, b]).$$

The "minimal spanning sets" introduced in *Subspaces* are linearly independent, as the next theorem shows.

Theorem 1. $v_1, \ldots, v_n$ *are linearly independent if and only if no v_i is a linear combination of the other v_j' s in the list.*

The function shrink (see page 109 of *Subspaces*) returns a minimal spanning set, which we now see is linearly independent. The next theorem is most easily remembered in this paraphrase "a spanning set is at least as big as an independent set."

Theorem 2. *Suppose $W = span(v_1, \ldots, v_n)$ and $w_1, \ldots, w_k \in W$ are linearly independent, then $k \leq n$.*

PROOF: We may assume that $v_1, \ldots, v_n$ are linearly independent by applying shrink. Let $A = [v_1, \ldots, v_n]$, then $\text{rank}(A) = n$. If we let $B = [w_1, \ldots, w_k]$, then $\text{rank}(B) = k$. Since for each $i = 1, \ldots, k$, $Ax = w_i$ has a solution, $\text{rank}([A, w_i]) = \text{rank}(A)$, and thus $\text{rank}(A) = \text{rank}([A, B])$. This means that

$$[A, B] \text{ reduces to } \begin{bmatrix} I_n & C \\ 0 & 0 \end{bmatrix}$$

where C is a $n \times k$ matrix. It follows that $k = \text{rank}(B) = \text{rank}(C) \leq n$. ∎

Bases

Let W be a subspace. A **basis** for W is a set of vectors $v_1, \ldots, v_n \in W$ such that

(1) $W = span(v_1, \ldots, v_n)$
(2) $v_1, \ldots, v_n$ are linearly independent.

By Theorem 1, a basis is the same as a minimal spanning set. If we take $W = \text{R}^n$, then a basis is given by $e_1, \ldots, e_n$. There are many other bases for R^n. We are now in a position to apply Theorem 2.

Theorem 3. *Suppose that $v_1, \ldots, v_n$ and $w_1, \ldots, w_k$ are two bases for W, then $n = k$.*

PROOF: By Theorem 2, since the v's span and the w's are independent, $k \leq n$. But the w's span and the v's are independent, so $n \leq k$. ∎

Dimension

In view of Theorem 3 we are in a position to make the following definition. The **dimension** of a subspace W can now be unambiguously defined as the number of elements in any basis for W. We will write $\dim(W)$ for this number. Since $e_1, \ldots, e_m$ is a basis for R^m, $\dim(\text{R}^m) = m$. The idea of dimension works well geometrically in R^3. A line through the origin is a subspace of dimension 1, a plane through the origin is a subspace of dimension 2, and all of 3–dimensional space, R^3, has dimension 3. The vector space $\{\vec{0}\}$ has no basis, since there is no independent set containing $\vec{0}$, and so has $\dim(W) = 0$.

We now consider the dimension of some of our familiar subspaces. To determine the dimension we need only find a basis.

In *Subspaces* we found a minimal spanning set for the row space by taking the nonzero rows of the reduced row echelon form. By Theorem 1, these nonzero rows are a linearly independent set which span the Row Space of A, and thus constitute a basis for the Row Space of A. The dimension of the Row Space of A is rank(A).

Recall B=shrink(A) see page 109 of *Subspaces* which finds a submatrix B of A with the same column space as A and the columns are linearly independent. The columns of B form a basis for the Column Space. The matrix B is obtained by removing the columns of A which do not become the columns of leading ones in the reduced row echelon form. There is a leading one for each nonzero row, so the number of vectors in the basis produced by shrink(A) is rank(A). The dimension of the Column Space of A is rank(A). The algorithm used to compute shrink(A) can be described as ***shrinking a spanning set down to a basis.***

It is also possible to ***enlarge an independent set to a basis.*** Suppose that $w_1, \ldots, w_n \in W$ is an independent set and suppose that $v_1, \ldots, v_m$ span W. We can enlarge $w_1, \ldots, w_n$ to a basis by observing that $w_1, \ldots, w_n, v_1, \ldots, v_m$ spans W and by shrinking to a basis of W. With a little care, we can make sure that $w_1, \ldots, w_n$ will be in the basis. In MATLAB

    A=rand(5,2)

    B=[A,eye(5)]

The columns of A are linearly independent and the columns of B span R^5. Thus shrink(B) is a basis for R^5 which includes the columns of A.

Theorem 4. *Let W be a subspace with dim(W) = n.*
(1) If $v_1, \ldots, v_n \in W$ are linearly independent, then

$$\mathrm{span}(v_1, \ldots, v_n) = W.$$

(2) If $\mathrm{span}(v_1, \ldots, v_n) = W$, then $v_1, \ldots, v_n$ are linearly independent.

In MATLAB take

    A=rand(5)

    rank(A)

Undoubtedly, the rank(A) is 5, which allows us to conclude that A row reduces to eye(5) and thus the columns of A are linearly independent. Since the dim(R^5) = 5, by Theorem 4 we conclude that the columns of A form a basis for R^5.

Theorem 5. *If $U \subseteq W$ are subspaces with dim(U) = dim(W), then $U = W$.*

Since the row space and the column space have the same dimension we get the following:

Theorem 6. $\mathrm{rank}(A) = \mathrm{rank}(A^T)$

The subspaces $V + W$ and $V \cap W$ (see page 108 of *Subspaces*) are closely related. One connection is provided by the dimension.

Theorem 7. *If V and W are subspaces, then*

$$dim(V + W) = dim(V) + dim(W) - dim(V \cap W)$$

PROOF: Suppose that the columns of A are a basis for $V \cap W$. Now let B be chosen so that the columns of $[A, B]$ are a basis for V and let C be chosen so that the columns of $[A, C]$ are a basis for W. The columns of $[A, B, C]$ are a linearly independent set which span $V + W$ and there are $dim(V) + dim(W) - dim(V \cap W)$ columns in $[A, B, C]$. ∎

Coordinates

If $v_1, \ldots, v_n$ is a basis for W, then for any $u \in W$ there exist unique scalars $r_1, \ldots, r_n$ such that

$$u = r_1 v_1 + \cdots + r_n v_n.$$

The scalars $r_1, \ldots, r_n$ are referred to as the ***coordinates of*** u ***with respect to*** $v_1, \ldots, v_n$. By Theorem 1 of *Subspaces*, finding the coordinates amounts to solvin g $Ax = u$, where $A = [v_1, \ldots, v_n]$.

PROBLEMS

1. Determine if the columns of the following matrices are linearly independent:

    (1) `magic(5)`
    (2) `list(5)`
    (3) `magic(6)`
    (4) `rand(2,5)`
    (5) `v*w where v=rand(5,1) and w=rand(1,5).`

See page 34 of *Building Matrices* for the definition of `list`.

2. Determine if the rows of the matrices in problem 1 are linearly independent.

3. Find a basis for the Column Space of each of the matrices in problem 1. Find a basis for the Row Space of each matrix in problem 1.

4. Let `A=magic(8)` and `B=A(:,[2,3,5])`. Show that the columns of B are a basis for the Column Space of A.

5. Let B be as in problem 4. Enlarge the columns of B into a basis for R^8.

6. Write a MATLAB function `E=enlarge(A)` which accepts a matrix A with linearly independent columns and returns a matrix E whose columns are a basis for R^m where A is $m \times n$ and E contains A as a submatrix.

7. Test your `enlarge` on the matrix B from problem 4.

8. Let V be the Column Space of `list(5)` and W be the Column Space of `A=v*w` where `v=rand(5,1)` and `w=rand(1,3)`. Determine $dim(V + W)$ and $dim(V \cap W)$.

9. What are the coordinates of `u=(1:5)'` with respect to the basis e_1, e_2, e_3, e_4, e_5? What are the coordinates of u with respect to the columns of `hilb(5)`? What are the coordinates of u with respect to the columns of `magic(5)`? Suppose that v is a vector whose coordinates with respect to the columns of `magic(5)` are `ones(5,1)`. What are the coordinates of v with respect to e_1, e_2, e_3, e_4, e_5? What are the coordinates of v with respect to the columns of `vander(rand(5,1))`? See page 32 of *Building Matrices* for `vander`.

10. Let a1=[cos(pi/4);sin(pi/4)] and

a2=[cos(3*pi/4);sin(3*pi/4)]

We are going to use a1,a2 as a basis for R^2 to graphically show the meaning of coordinates. Set the axes with axis([-2,2,0,2]) and use vecplot from *Graphics* to plot a1 and a2 (you can use hold to get both plotted on the same graph.) Let x=[.5;2] and use vecplot to plot x. Now compute the coordinates x1 and x2 of x with respect to a1 and a2. You should have x=x1*a1+x2*a2. Now plot x1*a1 and x2*a2. Finally complete the picture with vecplot(x,x1*a1) and vecplot(x2*a2,x). This worked out neatly since the basis vectors a1 and a2 were orthogonal to each other. Now take a1=[-.5;1] and a2=[1.5;2] and repeat the exercise with the same x.

COMMENTS

The rank of a matrix is difficult to compute. For example A=hilb(5)^3 is an invertible matrix since hilb(5) is invertible but both MATLAB's rank and rref tell us that the rank is 4. MATLAB's rank computes the rank using the SVD Decomposition. See *The Singular Value Decomposition* for a discussion of this.

Billiards

ABSTRACT
A simple game of billiards is to be played on an n-sided polygonal table. This illustrates the basics of vector geometry and parametic equations.

MATLAB COMMANDS
```
plot, rank, real, imag, \, +, *, -, '
```

LINEAR ALGEBRA CONCEPTS
Projection, Reflection

BACKGROUND
A billiards table is constructed by choosing n points $p_1, \ldots, p_n \in \mathbb{R}^2$. The table is bounded by the line segments p_1 to p_2, p_2 to p_3, $\ldots$, p_n to p_1. For simplicity, we want the points to have the following convexity property:

Each ray which originates inside (not on the boundary) of the region meets only one line segment on the boundary, unless it meets a vertex.

If we take an arbitrary set of points $p_1, \ldots, p_n$, we may have to reorder and delete some of the points to satisfy this property.

We will begin by choosing the points $p_1 = (0,0)$, $p_2 = (1,0)$, $p_3 = (2,2)$, and $p_4 = (0,1)$. In MATLAB type
```
x=[0 1 2 0 0];
y=[0 0 2 1 0];
plot(x,y)
hold
```
We start the game by picking a point, q, on the boundary, and a vector, v, pointing inside the region. We move a ball starting at q on the ray determined by v toward one of the edges of the region. When the ball hits the edge, it bounces according to the law of physics: *angle of reflection = angle of incidence.* Now the ball is speeding along a new vector towards another edge, and so on.

The essential mathematics is simple. The line segment between two points p_i and p_{i+1} is given by

$$\vec{x} = tp_i + (1-t)p_{i+1}$$

where t is a scalar parameter $0 \leq t \leq 1$. The ray with tail at q in the direction of the vector v is given by

$$\vec{x} = q + sv$$

where s is a scalar parameter $s \geq 0$. Let's choose the point $q = (.5, 0)$ and shoot it in the direction of $v = (0, 1)$. In MATLAB
```
q=[.5;0]; v=[0;1];
```

We can determine if, and where, the ray meets the segment by solving $q+sv = tp_i+(1-t)p_{i+1}$ for $s \geq 0$ and $0 \leq t \leq 1$. This is the same as $sv + t(p_{i+1} - p_i) = p_{i+1} - q$. The last equation leads to the 2×2 matrix equation

$$[v, \quad p_{i+1} - p_i] \begin{bmatrix} s \\ t \end{bmatrix} = p_{i+1} - q$$

There are three possibilities for this matrix equation:

(1) The ray meets the line segment with $s > 0$ and $0 \leq t \leq 1$. This would happen in our example with p_3 and p_4. In MATLAB we first check that there is a solution and then find it.

```
A=[v,[x(4)-x(3);y(4)-y(3)]],
rank(A),
b=[x(4)-q(1);y(4)-q(2)]
u=A\b
```

Notice that u(1)= $s > 0$ and $0 \leq$ u(2)= $t \leq 1$.

(2) The line through the ray meets the line through the segment, but the ray does not hit the segment: $s < 0$ or $t > 1$ or $t < 0$. In our example this would happen with p_2 and p_3.

```
A=[v,[x(3)-x(2);y(3)-y(2)]],
rank(A),
b=[x(3)-q(1);y(3)-q(2)]
u=A\b,
```

Here we notice that u(1)= $s < 0$ and u(2)= $t < 0$.

(3) The ray is parallel to the segment. In this case the coefficient matrix is singular. In our example this happens with p_1 and p_4.

```
A=[v,[x(1)-x(4);y(1)-y(4)]],
rank(A)
```

Once it has been determined that the ray hits the segment, then the point of intersection is obtained immediately from either parameter s or t. Now the ball is to carom off the side and we need to know the new direction vector. This is obtained easily by reflecting the vector $-v$ through a vector perpendicular to the edge The minus sign on $-v$ insures that the vector is going in the correct direction after the collision. The direction of the edge vector is $w = p_i - p_{i-1}$. From page 44 of *Graphics* we know that the reflection of v perpendicular to w is given by

$$u = v - 2\text{proj}_w(v)$$

where

$$\text{proj}_w(v) = (\frac{v \cdot w}{w \cdot w})w$$

When we reflect $-v$ we get

$$-v + 2(\frac{v \cdot w}{w \cdot w})w$$

Returning to our example we compute the point of intersection and the reflected vector and plot them.

```
A=[v,[x(4)-x(3);y(4)-y(3)]],
b=[x(4)-q(1);y(4)-q(2)]
u=A\b;
p=q+u(1)*v;
w=[x(4)-x(3);y(4)-y(3)];
w=-v+2*(v'*w)/(w'*w)*w;
plot([p(1);q(1)],[p(2);q(2)]);
```

We can put all of this information in a simple program.

```
function [p,w,t]=nextpt(q,v,x,y)

% [p,w,t]=nextpt(q,v,x,y). On a billiard
% table determined by "x" and "y" with the
% ball placed at "q" with direction "v"
% this will compute the side of the table
% that the ball will hit and the point "p"
% together with the new direction "w."
% If "t" is either 0 or 1, then the ball
% has hit a vertex.

% First locate the side where it hits.
n=length(x);
s=0; t=.5;
i=1;
while s==0 & i<n,
    i=i+1;
    A=[v,[x(i)-x(i-1);y(i)-y(i-1)]];
    % Check if system is consistent
    if rank(A)==2
        w=A\([x(i);y(i)]-q);
        %the ball properly meets the segment
        if w(1)>eps & w(2)>eps & w(2)<1-eps,
            s=w(1);
            t=w(2);
        end
        % the ball hits a vertex
        if w(1)>eps
            if abs(w(2))<eps | abs(w(2)-1)<eps
                s=w(1);
                t=w(2);
```

```
            end
        end
    end
end

% Compute the new point and vector.
p=q+s*v;
w=[x(i)-x(i-1);y(i)-y(i-1)];
w=-v+2*v'*w/(w'*w)*w;
```

We can watch the ball bounce around the table with the following program.

```
function billiard(q,v,x,y)

% billiard(q,v,x,y). Plays a
% game of billiards on a table
% determined by "x" and "y"
% in the direction "v." To stop
% type "control c."

plot(x,y);
hold

t=.5;
while t > 10*eps & abs(t-1) > 10*eps,
    [p,w,t]=nextpt(q,v,x,y);
    plot([p(1);q(1)],[p(2);q(2)]);
    pause(1);
    q=p;
    v=w;
end
```

Try this on our table with q=[0;.5] and v=[1;1].

PROBLEMS

1. Run the billiard function on some other starting positions. Try q=[0;1] and v=[.5;-1] and q=[.5;0] and v=[0;1].

2. Set up another table
    x=[1,2,3,3,2,1], y=[1,2,1,0,0,1].
Try q=[2;0], v=[1;.1], v=[1;.2], and v=[1;.5].

3. We can make this into a game of ***pocket billiards*** by halting play when the ball goes near a vertex. This is most easily done by computing norm(p-[x(i);y(i)]) and norm(p-[x(i-1);y(i-1)] since norm is the same as the Euclidean distance function. Write a modification of `billiard` called `pocket(q,v,x,y,r)` which halts if the ball is within `r` of a vertex.

4. Play `pocket` on the table in the Background discussion using `r=.1`.

5. Play `pocket` on the table in problem 2 using `r=.1`.

6. The object of ***m-cushion billiards*** is to hit *m* sides of the table and then go into a pocket. You may hit a side repeatedly. Play a game of 4-cushion billiards on the table in the Background discussion starting at `q=[0;0]`. Use a pocket size of `r=.1`.

7. Play a game of 5-cushion billiards on the table in problem 2 starting at `q=[2;0]` using a pocket size of `r=.1`.

8. Set up a regular hexagon for a table and try to win a game of 6-cushion billiards. This can be done with `t=[0:6,0]`, `x=cos(2*pi*t/6)`, `y=sin(2*pi*t/6)`, `plot(x,y)`.

9. Modify the program `billiard` so that it will work with a table which does not satisfy the convexity condition. Try it on the following table:

    `x=[4, 0, 0, -2, -2, 0, 0, 4]`
    `y=[0, 3, 1, 1, -1, -1, -3, 0]`.

10. Another modification of the `billiard` program is to make a game of ***bumper bowling***. We can set up a lane with

    `x=[0,10,10,0,0];`
    `y=[0,0,2,2,0];`

Now we bowl from the starting point of `q=[0;1]`. After the ball bounces off the sides it crashes into the back wall determined by the segment from `[10;0]` to `[10;2]`. The objective is to hit the lead pin which sits somewhere on the back wall. Modify `billiard` to make a MATLAB function `bumper(v,pin,rad)` which plays this game. If you come within `rad` of the point `pin` which sits on the back wall, the display "YOU WIN!" appears, otherwise "SORRY LOSER" is displayed. See how many times you can get the ball to bounce before hitting `pin`.

The Null Space

ABSTRACT

The set of solutions to the equation $Ax = \vec{0}$ is a subspace, called the Null Space of A. We find a basis for this subspace.

MATLAB COMMANDS

```
null, rref, sum, -, *
```

LINEAR ALGEBRA CONCEPTS

Reduce Row Echelon Form, Subspace, Null Space, Basis, Rank

BACKGROUND

Let A be an $m \times n$ matrix. ***The Null Space of*** A is defined to be

$$\{v \in \mathrm{R}^n : Av = \vec{0}\}.$$

Theorem 1. *The Null Space of A is a subspace.*

Suppose that we have solutions x_1 and x_2 to the nonhomogeneous equation $Ax = b$. Then $Ax_1 = b = Ax_2$, so that $Ax_1 - Ax_2 = A(x_1 - x_2) = \vec{0}$. Thus $x_1 - x_2$ is in the Null Space of A. That is there is a z in the Null Space of A such that $x_1 - x_2 = z$ which gives $x_1 = x_2 + z$. This last equation tells us that if we know just one solution x_2 to the equation $Ax = b$, then all other solutions have the form $x_2 + z$ where z is in the Null Space of A.

We are going to compute a basis for the Null Space. The process is given in problem 10 of *Systems of Linear Equations.* In MATLAB try

```
A=magic(8)
```

We want the reduced row echelon form.

```
A=rref(A)
```

Find the columns which contain the leading 1's and the columns which correspond to the free variables. The function `lead` appears on page 27 of *Systems of Linear Equations*

```
L=lead(A)'; F=~L;
```

Watch that transpose! Compute the rank.

```
r=sum(L); s=sum(F);
```

Notice that `s=n-r`. Here n=8. Throw out any rows of zeros in `A`.

```
A=A(1:r,:)
```

At this stage we can generate arbitrary solutions to `Ax=b` by assigning

```
x=zeros(size(L));
x(F)=rand(s,1)
x(L)=-A*x
```

You can quickly check this with

```
A*x
```

We want to get a basis $v_1, \ldots, v_s$ for the Null Space of A. To do this we will assign x(F)=e_i where e_i is the ith standard basis vector.

```
E=eye(s);
x=zeros(size(L)); x(F)=E(:,1); x(L)=-A*x; B=x;
x=zeros(size(L)); x(F)=E(:,2); x(L)=-A*x; B=[B,x];
x=zeros(size(L)); x(F)=E(:,3); x(L)=-A*x; B=[B,x];
x=zeros(size(L)); x(F)=E(:,4); x(L)=-A*x; B=[B,x];
x=zeros(size(L)); x(F)=E(:,5); x(L)=-A*x; B=[B,x];
```

Do not be surprised to see

```
B(F,:)
```

Now the matrix B contains s=n-r solutions. By the next theorem, the columns of B form a basis for the Null Space of A. Thus the Column Space of B is the Null Space of A.

Theorem 2. *Let A be an $m \times n$ matrix, the dimension of the Null Space of A is*

$$n - rank(A).$$

It is frequently useful to be able to switch back and forth between the Null Space representation and the Column Space representation of a subspace. In this line we ask, given a matrix A is there a matrix B where the Null Space of B= Column Space of A? This question will be answered in *Projections*.

Theorem 3. *Let A be any matrix, then*
$$\text{Null Space of } A = \text{Null Space of } A^T A$$
It follows that if s is the dimension of the Null Space of either A or $A^T A$, then rank$(A) =$ $n - s = rank(A^T A)$.

Suppose that V is the Null Space of A and W is the Null Space of B. Then we can represent $V \cap W$ as the Null Space of $[A; B]$, since $Ax = \vec{0}$ and $Bx = \vec{0}$ if and only if $[A; B]x = \vec{0}$. See page 108 for the definition of $V \cap W$.

MATLAB has a function B=null(A) which returns an orthonormal basis for the Null Space of A. Consult *Orthonormal Bases* for information about orthonormal bases. An orthonormal basis for the null space can be computed using the singular value decomposition (see *The Singular Value Decomposition*.) MATLAB uses the QR Decomposition.

PROBLEMS

1. Compute the dimension of the Null Space and use null to find a basis for the Null Space of each of the following matrices (see *Building Matrices* for the definition of these matrices):

(1) list(6)

    (2) `rand(5)`
    (3) `vander((1:5)')`
    (4) `compan([1,1,1])`
    (5) `A=magic(8); A=A(1:4,:)`

2. Use the method described in the Background to find a basis for the Null Space for each of the matrices in problem 1. You will notice that you do not get the same matrix as `null` returns.

3. Write a MATLAB function `B=nullbase(A)` which returns a matrix B whose columns form a basis for the Null Space of A using the method described in the Background.

4. Test your `nullbase` on the examples in problem 1 comparing your answers with problem 2.

5. Suppose that V = Null Space of A for `A=list(8)` and W = Null Space of B for `B=magic(8)`. Compute $\dim(V \cap W)$ and $\dim(V + W)$. Find a basis for $V \cap W$ using the method described on page 121 or if you have completed problem 3, use `nullbase`.

6. Let `A=magic(8); A=A(1:4,:)`, then let `B=A'*A`. Compute `rank(A)` and `rank(B)`. Now compute bases for the Null Space of A and the Null Space of B using both `null` and `nullbase`.

7. Let `A=rand(4,6)` and compute a basis for the Null Space of A, `B=null(A)`. Show that the Column Space of B = Null Space of A. Why should this not be a surprise?

8. Let A be the matrix in problem 6 and let `b=ones(4,1)`. Find v and B so that all solutions of `Ax=b` have the form `v+B*w` where v is a solution to `Ax=b` and B is a basis for the Null Space of A.

9. Write a MATLAB function `[v,B]=nonhom(A,b)` which returns a n $\times$ 1 vector v and an n$\times$(n$-$rank(A)) matrix B such that if `Ax=b` is consistent, then every solution to `Ax=b` has the form `v+Bw` for some vector w. Test your `nonhom` on the matrices in problem 8.

Projections

ABSTRACT

We define the projection of a vector onto a subspace. To make this definition, a new subspace, the orthogonal complement, is introduced. Given a subspace W, the orthogonal complement is the subspace of vectors which are perpendicular to all vectors in W.

MATLAB COMMANDS

```
null, plot, norm, orth, *, ', \
```

LINEAR ALGEBRA CONCEPTS

Column Space, Null Space, Projection, Dimension, Orthogonal Complement

BACKGROUND

If W is a subspace of R^m, then we define the **orthogonal complement of W** to be the set

$$W^\perp = \{v \in \mathrm{R}^m : v \cdot w = 0 \text{ for all } w \in W\}.$$

Read this as "W perp," for perpendicular. This is the set of all vectors which are perpendicular to all vectors in W. For a concrete example, consider a plane in 3-dimensional space, the orthogonal complement of the plane is the line perpendicular to the plane and passing through the origin.

Theorem 1. $W^\perp$ *is a subspace of* R^m.

Note that if $w \in W \cap W^\perp$, then $w \cdot w = 0$ and so $w = \vec{0}$. We will now tie the orthogonal complement to some familiar subspaces.

Theorem 2. *Given any matrix A*

$$(\textit{Column Space of } A)^\perp = \textit{Null Space of } A^T.$$

PROOF: If v is in the perp of the Column Space of A, then $v \cdot v_i = 0$ for all columns v_i of A. Thus $v_i^T v = 0$ for all columns v_i of A and so $A^T v = \vec{0}$. It follows that v is in the Null Space of A.

Now suppose that $A^T v = \vec{0}$, so that $v^T v_i = 0$ for all columns v_i of A. Let $w = Az$ be a typical element of the Column Space of A. Then

$$v^T w = v^T Az = [v^T v_1, \ldots, v^T v_n]z = \vec{0} \cdot z = 0$$

and so v is perpendicular to w for all w in the Column Space of A, showing that v is in $W^\perp$. ∎

Every vector v in R^m can be decomposed as $v = w + u$ where $w \in W$ and $u \in W^\perp$. We define w to be **the projection of v onto W** and u to be the **projection of v onto $W^\perp$** or the **projection of v perpendicular to W.**

Theorem 3. *If $v \in \mathbb{R}^m$, then there exist unique vectors $w \in W$ and $u \in W^\perp$ such that $v = w + u$.*

PROOF: Let A be an $m \times n$ matrix where the Column Space of $A = W$. By shrinking, (see page 112) we may assume that rank$(A) = n$. The matrix $A^T A$ is $n \times n$ and has rank n (see Theorem 3 of *The Null Space*.) Thus $A^T A$ is invertible, so we can define

$$x = (A^T A)^{-1} A^T v \quad \text{and} \quad w = Ax.$$

Notice that w is in W = the Column Space of A, and also that x is the solution to the *normal equations*

$$A^T A x = A^T v.$$

Thus $A^T w = A^T v$, and so $A^T(w - v) = \vec{0}$. It follows that $u = w - v$ is in the Null Space of A^T.

To show uniqueness suppose that $v = w_1 + u_1 = w_2 + v_2$ where $w_1, w_2 \in W$ and $u_1, u_2 \in W^\perp$. Then $w_1 - w_2 = v_2 - v_1 \in W \cap W^\perp$. By an earlier remark, $w_1 - w_2 = \vec{0}$ so $w_1 = w_2$ and similarly, $u_1 = u_2$. ∎

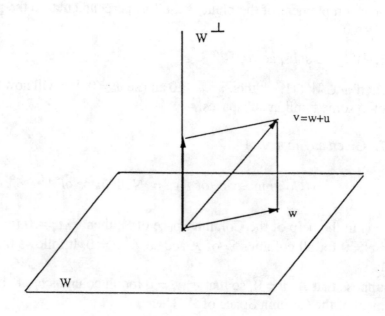

The proof of the theorem provides explicit formulas for w and u.

$$\text{proj}(A, v) = w = A(A^T A)^{-1} A^T v$$

and $u = v - w$ when rank$(A) = n$. Special cases of these formulas which apply to vectors appeared in *Graphics*. In the case of $A = x$, a column vector, we get the *projection of v onto x* (recall from page 44)

$$\text{proj}(x, v) = \frac{x \cdot v}{x \cdot x} x$$

Suppose that we want to project the vector $(1, 1, 1)$ onto the plane $x + 2y - z = 0$. We notice that this is the Null Space of B where $B = [1, 2, -1]$. We first need to get a matrix A whose column space is the same as the Null Space of B. In MATLAB

```
A=null(B)
```

Now solve the normal equations

```
v=[1; 1; 1]
x=(A'*A)\ (A'*v)
```

The projection is given by

```
w=A*x
```

We can check our work by showing that w and v-w are orthogonal.

```
(v-w)'*w
```

Our approach to projections has been based on a decomposition of the space R^n. We will try to make our definition of the projection more convincing by showing that $\mathrm{proj}(A, v)$ is the vector in the Column Space of A which is closest to v. To do this we need the assistance of an old friend, Pythagoras. Recall that $\|x\| = \sqrt{x_1^2 + \cdots + x_n^2}$.

Theorem 4. *The Pythagorean Theorem. Suppose that $x \cdot y = 0$, then*

$$\|x + y\|^2 = \|x\|^2 + \|y\|^2.$$

PROOF:

$$
\begin{aligned}
\|x + y\|^2 &= \sum (x_i + y_i)^2 \\
&= \sum (x_i^2 + 2x_i y_i + y_i^2) \\
&= \left(\sum x_i^2\right) + 2x \cdot y + \left(\sum y_i^2\right) \\
&= \|x\|^2 + \|y\|^2.
\end{aligned}
$$

Theorem 5. *Let w be the projection of v onto the subspace W. Then*

$$\|v - w\| \leq \|v - z\| \text{ for all } z \in W.$$

PROOF: Write $v = w + u$ where $w \in W$ is the projection and $u = v - w \in W^{\perp}$. Now if z is any element of W, then $v - z = (v - w) + (w - z)$ and since $v - w = u \in W^{\perp}$ and $w - z \in W$ we have $(v - w) \cdot (w - z) = 0$. By the Pythagorean Theorem $\|v - z\|^2 = \|v - w\|^2 + \|w - z\|^2 \geq \|v - w\|^2.$ ∎

In problem 8 there is a MATLAB computation to illustrate that the projection is the vector in the subspace which is closest to v.

Notice that since $W^\perp$ = the Null Space of A^T, $\dim(W^\perp) = m - \text{rank}(A^T) = m - \text{rank}(A)$. Also since $\dim(W) = \text{rank}(A)$, we get

$$m = \dim(W) + \dim(W^\perp).$$

Theorem 6. $(W^\perp)^\perp = W.$

PROOF: Let $w \in W$, so $u \cdot w = 0$ for all $u \in W^\perp$. But this is the definition of $(W^\perp)^\perp$, *i.e.*, $u \in (W^\perp)^\perp$. We have shown that $W \subseteq (W^\perp)^\perp$. Now $\dim((W^\perp)^\perp) = m - \dim(W^\perp) = \dim(W)$, and so $W = (W^\perp)^\perp$ (see Theorem 5 of *Dimension and Rank*). ∎

By Theorem 2, if W = Column Space of A, then $W^\perp$ = Null Space of A^T. By Theorem 4, $W = (W^\perp)^\perp = $(Null Space of $A^T)^\perp$. This result is called the **Fredholm Alternative:**

$$\text{Column Space of } A = (\text{Null Space of } A^T)^\perp$$

As an application of the Fredholm Alternative consider the question we raised in *The Null Space:* given a matrix A can we find a matrix B such that

$$\text{the Null Space of } B = \text{Column Space of } A?$$

From the Fredholm Alternative we know that

$$\text{the Column Space of } A = (\text{Null Space of } A^T)^\perp.$$

Now find a matrix B whose columns are a basis for the Null Space of A^T, so that

$$\text{the Column Space of } A = (\text{Column Space of } B)^\perp,$$

and, applying Theorem 2,

$$\text{Column Space of } A = (\text{Null Space of } B^T).$$

In MATLAB

```
A=rand(5,2)
```

Find a basis for the Null Space of A'

```
B=null(A')
```

Complete the process with the transpose of B

```
B=B'
```

Check with

```
B*A
```

PROBLEMS

1. For A=magic(8) find a matrix B where the Null Space of B is the perp of the Column Space of A. Now find a matrix C where the Column Space of C is the perp of the Column Space of A. Repeat this for A=list(6). See page 34 of *Building Matrices* for list.

2. Let A=magic(8); A=A(:,1:3) and let W be the Column Space of A. For v=(1:8)' compute the projection of v onto the Column Space of A. What is the projection of v perpendicular to the Column Space of A?

3. Let A=list(6) and let W be the Column Space of A. For v=[1,0,1,0,1,0] compute the projection of v onto W and the projection of v perpendicular to W.

4. Let A=list(6) and let W be the Null Space of A. For v=ones(6,1) compute the projection of v onto W and the projection of v perpendicular to W.

5. Write a MATLAB function w=proj(A,v) which computes the projection of v onto the Column Space of A. Your program should allow the possibility that the columns of A are not linearly independent. This can be handled with the MATLAB function orth (see page 165) or shrink (see page 109)

6. Check your function proj against the outcomes of problems 2 and 3.

7. Let A=[1;2], v=[1;1], and let w be the projection of v onto the Column Space of A, and u the projection of v perpendicular to the Column Space of A. Using vecplot from page 43 of *Graphics* plot these four vectors. Check that w and u really are perpendicular by computing w'*u. Now repeat this with v=[-1;1].

8. The projection of a vector v onto the Column Space of A produces the vector w in the Column Space of A which is closest to the vector v by Theorem 5. We are going to set up an exercise to illustrate this. Let

$$A = \begin{bmatrix} 1 & 1 \\ 0 & 1 \\ 2 & 2 \end{bmatrix} \qquad b = \begin{bmatrix} 1 \\ 0 \\ 1 \end{bmatrix}$$

The Column Space of A is a plane and b is a vector which is not on the plane. Now let $w = \text{proj}(A, b)$ so that w is a point on the plane. We measure the distance from w to b using the MATLAB function norm

$$\text{norm}(w - b) = \sqrt{(w(1) - b(1))^2 + (w(2) - b(2))^2 + (w(3) - b(3))^2}.$$

Let x=rand(2,1); u=A*x, so that u is a point on the plane and look at the parametric form of the line segment from u to w

$$t\text{w} + (1 - t)\text{u} \qquad 0 \le t \le 1$$

Look at points on the line given by t=0:.05:1 and compute the distances

```
    for i=t
        norm(i*w+(1-i)*u -b)
    end
```

At what point does the minimum value occur?

9. In problem 8 the equation $Ax = b$ has no solution, so b is not in the Column Space of A. By the Fredholm Alternative, b is not in the perp of the Null Space of A^T. Thus there is a u such that $A^T u = \vec{0}$ and $u \cdot b \ne 0$. Find such a vector u.

10. The matrix $P = A(A^T A)^{-1} A^T$ is the matrix used to project onto the Column Space of A. This matrix has the interesting feature that $P^2 = P$. Can you see why this is true by thinking about the projection of the projection of a vector? Let A=rand(7,4), form P, and look at P*P-P. What happens to A*inv(A'*A)*A' when rank(A)=n and A is square? Compare rank(A) and rank(P).

11. MATLAB has a function B=orth(A) which returns an orthogonal matrix B with the same column space as A. See *Orthonormal Bases* for the definition of orthonormal. When the columns are orthonormal the projection formula simplifies to w=B*B'*v. In the example in problem 8, let B=orth(A) and form the projection w=B*B'*b. Does the answer agree with the answer in problem 8?

Least Squares

ABSTRACT

A system $Ax = b$ may not have a solution. When this happens we choose the best available answer by minimizing the residual error. Other times a system may have many solutions and we need to select a particular solution. In this case we choose the shortest solution. These are both instances of least squares optimization. Many concepts from *Projections* are used in this project.

MATLAB COMMANDS

```
norm, plot, *, ', \
```

LINEAR ALGEBRA CONCEPTS

Systems of Linear Equations, Projections, Norm

BACKGROUND

Underdetermined Least Squares

Let us consider a system $Ax = b$ where A is $m \times n$ which may or may not have a solution. If x is any vector we can form

$$r = abs(Ax - b).$$

r is called the ***residual vector*** and the ***residual*** is $\|r\|_2$. The residual is a measure of the error in the "solution" x. If $r = (r_1, \ldots, r_n)$, then the MATLAB function $\text{norm}(r, 2) = \|r\|_2 = \sqrt{r_1^2 + \ldots + r_n^2}$. The ***least squares solution*** to $Ax = b$ is the vector x where $\|r\|_2$ is minimal. Notice that

The least squares solution is not a solution to the system $Ax = b$ when $Ax = b$ is inconsistent; it is the vector x with the smallest residual.

We are going to derive a formula for the least squares solution. Theorem 5 from *Projections* tells how to find the least squares solution.

Theorem 1. *Let w be the projection of b onto the Column Space of A. Then*

$$\|b - w\|_2 \leq \|b - Ax\|_2 \text{ for all } x \in R^n.$$

Thus the least squares solution to $Ax = b$ is a vector x where $Ax = \text{proj}(A, b)$.

An $m \times n$ system where $m > n$ is a system which is likely to be inconsistent and is called ***overdetermined***. The least squares solution is as much as one could expect. By Theorem 1, we see that the least squares solution is found by solving $Ax = \text{proj}(A, b)$. From *Projections* we know that $\text{proj}(A, b) = Ax$ where x is found by solving the normal equations (see page 124 of *Projections*) $A^T Ax = A^T b$. Thus, stated directly, the least squares solution is the solution to the normal equations, $A^T Ax = A^T b$. When $\text{rank}(A) = n$, $A^T A$ is invertible (see Theorem 3 of *The Null Space*) and thus there is a unique solution, $x = (A^T A)^{-1} A^T b$. MATLAB's \ will return the least squares solution to an overdetermined system.

Regression Lines

We will now consider how to approximate some data points $(a_1, b_1), \ldots, (a_m, b_m)$ with a line. Two points will determine a line, but three or more points are unlikely to be collinear, and so finding a line through these points is usually impossible. This is a perfect problem for a least squares solution.

Suppose the line is to have the equation $y = c_1 x + c_2$ where c_1 and c_2 are unknowns. Then we get the following $m \times 2$ system

$$c_1 a_1 + c_2 = b_1$$
$$c_1 a_2 + c_2 = b_2$$
$$\vdots$$
$$c_1 a_m + c_2 = b_m.$$

Giving rise to the $m \times 2$ matrix equation

$$\begin{bmatrix} 1 & a_1 \\ 1 & a_2 \\ \vdots & \vdots \\ 1 & a_m \end{bmatrix} \begin{bmatrix} c_2 \\ c_1 \end{bmatrix} = \begin{bmatrix} b_1 \\ b_2 \\ \vdots \\ b_m \end{bmatrix}$$

Typically, this equation will not have a solution and so we will appeal to the least squares solution. The resulting line is the ***least squares regression line.*** These are easy to compute. Consider the data given by (a(i),b(i)),

```
a=[0 1 2 3]'; b=[0 1 .5 1]';
A=[a,ones(4,1)];
c=(A'*A)\(A'*b);
```

Plot the points and the line. The function `polyplot` appears on page 42 of *Graphics*. The polynomial needs to be a row vector for plotting with `polyplot`.

```
plot(a,b,'*'), hold
polyplot(0,3,c')
```

Now suppose that we want to approximate the data $(a_1, b_1), \ldots, (a_m, b_m)$, with a polynomial of degree $n - 1$,

$$c_1 x^{n-1} + c_2 x^{n-2} + \ldots + c_{n-1} x + c_n.$$

As before, we substitute the data into the polynomial to get the equations

$$c_1 a_1^{n-1} + c_2 a_1^{n-2} + \ldots + c_{n-1} a_1 + c_n = b_1$$
$$c_1 a_2^{n-1} + c_2 a_2^{n-2} + \ldots + c_{n-1} a_2 + c_n = b_2$$
$$\vdots$$
$$c_1 a_m^{n-1} + c_2 a_m^{n-2} + \ldots + c_{n-1} a_m + c_n = b_m$$

Which gives rise to the matrix equation

$$\begin{bmatrix} 1 & a_1 & \cdots & a_1^{n-2} & a_1^{n-1} \\ 1 & a_2 & \cdots & a_2^{n-2} & a_2^{n-1} \\ \vdots & \vdots & & \vdots & \vdots \\ 1 & a_m & \cdots & a_m^{n-2} & a_m^{n-1} \end{bmatrix} \begin{bmatrix} c_1 \\ c_2 \\ \vdots \\ c_n \end{bmatrix} = \begin{bmatrix} b_1 \\ b_2 \\ \vdots \\ b_m \end{bmatrix}$$

Notice the similarity with the Vandermonde matrix (see page 32 of *Building Matrices* .) This matrix is not necessarily square. As before, to get the least squares solution, project b onto the Column Space of the coefficient matrix. Take the data (a,b) from above and fit a least squares cubic polynomial to it.

```
A=[a.^2,a,ones(a)]
c=(A'*A)\(A'*b);
polyplot(0,3,c')
```

The MATLAB function `polyfit` will find the least squares polynomial to a data set. Try it on this data.

```
c=polyfit(a,b,3)
```

There is no mathematical reason for choosing one particular degree of polynomial over another. Assuming that the data is related to some scientific theory, that theory may suggest the appropriate degree of approximation. You should also be aware that the numerical problem becomes more ill conditioned as you increase the degree.

Underdetermined Least Squares

The **underdetermined** case occurs when $m < n$. Here there are usually many solutions to $Ax = b$ and we need to select one. In this case, we select the solution x where $\|x\|_2$ is the smallest. This differs from the overdetermined least squares problem where we choose the x with the smallest residual. We will give the general least squares solution to this problem.

Theorem 3. *Suppose that x is the least squares solution to the underdetermined system $Ax = b$ and $rank(A) = m$, then*

$$x = A^T(AA^T)^{-1}b.$$

PROOF: Suppose that x and y are both solutions to $Ax = b$, then $A(y - x) = Ay - Ax = b - b = \vec{0}$, so that $y - x$ is in the Null Space of A. Letting $z = y - x$ we see that $y = x + z$ where z is from the Null Space of A. Now if we were to choose x to be perpendicular to z, then by the Pythagorean Theorem, $x \cdot z = 0$ and $\|y\|^2 = \|x + z\|^2 = \|x\|^2 + \|z\|^2 \geq \|x\|^2$. Thus if we can get $x \cdot z = 0$ for all z in the Null Space of A, we would have the solution with the minimal 2-norm.

Here is how to get x perpendicular to z. Let x come from the perp of the Null Space, which by the Fredholm Alternative (see page 126 of *Projections*) is the Column Space of A^T. Thus we need to solve $A^T u = x$ and $Ax = b$. Putting this together, $AA^T u = b$, which has a unique solution as long as $rank(A^T) = rank(A) = m$. Thus $u = (AA^T)^{-1}b$ and $x = A^T u = A^T(AA^T)^{-1}b$. ∎

PROBLEMS

1. Find the least squares polynomials of degrees $n = 1, 2, 3, 4$ for the data

$$x = [-2, -.5, 0, .5, 1]^T \qquad y = [0, -1, 0, 2, 0]^T$$

Using the `polyplot`, graph these polynomials and the data points. Compute the 2-norm of the residual for each of the solution vectors.

2. A standard Calculus problem is to find the point on the plane $x + y + z = 2$ which is closest to the origin. This is generally set up as a problem to minimize the function $f(x, y, z) = \sqrt{x^2 + y^2 + z^2}$ subject to the constraint $x + y + z = 2$. Formulate this as an underdetermined least squares problem and solve it with MATLAB.

3. Consider the MATLAB function A*B for $n \times n$ matrices A and B. MATLAB will keep track of the number of arithmetic operations needed to perform an operation with the command `flops` (see *Flops.*) Try this:

```
A=rand(5); B=rand(5);
flops(0)
A*B;
flops
```

You should get 250. Use the pair $(5, 250)$ as a data point. Now let

```
x=1:10;
for i=x,A=rand(i);B=rand(i);flops(0);A*B;y=[y,flops];end
plot(x,y,'*'), hold
```

This gives you a nice data set. Using least squares find the least squares regression line for this data.

```
c=polyfit(x,y,1);
polyplot(0,10,c),
```

Now try the quadratic and cubic. In *Flops* it is shown that this is a cubic polynomial. You can see the relative error with the least squares approximation

```
d=polyval(c,x);
abs(y-d)./abs(y)
```

4. Try the experiment from problem 3 on A\b instead of A*B. Since this operation is based on Gaussian Elimination we should be close to a cubic polynomial in n where A is $n \times n$. Get 10 pieces of data as in problem 3 and see if you can get a good fit. Again plot and check the relative error. How well does your approximating polynomial work when you use A=triu(rand(n))?

5. We should be able to get `etime` (see *Flops*) results similar to the `flops` results in problem 3. Here again we expect the polynomial should be a cubic. Since clock times are more erratic than flop counts, I suggest that you get an average for each value of n as in problem 7 of *Flops*. On some of the faster machines the `etime` values for the lower values of n may be too small to work with. In that case use n=20:30 or possibly larger.

6. Construct the following data:

```
p=[1,-8,28,-56,70,-56,28,-8,1]
x=.9:.004:1.1
```

Now take y=polyval(p,x). To get the picture plot(x,y). Now run polyfit on this data to get a polynomial of degree 8 to fit the data. You can see how close it comes by looking at a plot of the least squares polynomial. Now try x=.95:.002:1.05 and repeat the same process. The graph should look good, but the coefficients are off a bit. Now try x=.97:.0008:1.03. You will see significant round-off in the plot but look at how nicely the least squares polynomial fits the data.

7. Write a MATLAB function c=lstsq(A,b) which finds the least squares solution to an overdetermined system Ax=b. Your function should be based on solving the normal equations A'*A*x=A'*b using either MATLAB's inv or \. You may assume that rank(A)=n so that A is invertible. Check your lstsq on the data in problem 1.

8. Collect data from the function logistic (see page 73 of *Chaos and Fractals*.) To do this set ax=[0,200,0,1] and run y=logistic(3,ax,200) plot and hold it. Now find the least squares line through the data x=1:200 and y. Using polyplot plot the least squares line through this data.

9. A *line spline* is function $S(x)$ which is built from pieces of linear functions $p_i(x)$ and points $x_1 < x_2 \ldots < x_n$ such that

$$S(x) = \begin{cases} p_1(x) & \text{if } x_1 \leq x \leq x_2 \\ p_2(x) & \text{if } x_2 \leq x \leq x_3 \\ \vdots & \vdots \\ p_{n-1}(x) & \text{if } x_{n-1} \leq x \leq x_n \end{cases}$$

Given a set of points $(a_1, b_1), \ldots, (a_n, b_n)$ it is a simple matter to get a line spline through these points simply define $x_i = a_i$ for $i = 1, \ldots, n - 1$ and let

$$p_i(x) = \left(\frac{b_{i+1} - b_i}{a_{i+1} - a_i} \right) (x - a_{i+1}) + b_{i+1}$$

Find the polynomials $p_i(x)$ for the data a=0:5 and b=[1,0,2,4,3,-1]. Graph the spline.

10. Returning to the line splines introduced in problem 9. Suppose that the spline knots $x_1, \ldots, x_m$ are given independently of the data, $(a_1, b_1), \ldots, (a_n, b_n)$. We cannot insure that the line spline will now pass through the data points, so we are led to find a least squares solution. This is done as follows: Let $R_i(x) = \frac{x - x_i}{x_{i+1} - x_i}$ and $L_i(x) = \frac{x_{i+1} - x}{x_{i+1} - x_i}$ be the linear Lagrange polynomials. $p_i(x) = c_i L_i(x) + c_{i+1} R_i(x)$ where c_i and c_{i+1} are coefficients. We must determine the coefficients $c_1, \ldots, c_m$. Notice that we require $p_i(x_{i+1}) = c_{i+1}$ and $p_{i+1}(x_{i+1}) = c_{i+1}$ so that the lines will touch at the endpoints. For each i where $1 \leq i \leq m - 1$ there is an interval $[x_i, x_{i+1}]$ which contains some of the points a_j. We get the equation

$$p_i(a_j) = c_i L_i(a_j) + c_{i+1} R_i(a_j) = b_j$$

when $x_i \leq a_j \leq x_{i+1}$. In all we get n equations, one for each pair (a_j, b_j), but which equation we get depends on which interval a_j falls in. If A is the $n \times m$ matrix of coefficients, c is the $m \times 1$ matrix of unknown coefficients, and b is the $n \times 1$ matrix of the b_j's, then we have an overdetermined least squares problem. Using the data from problem 9 and letting $x_1 = 0$, $x_2 = 2$, $x_3 = 4$, $x_4 = 5$ find the least squares line spline and plot it with the data points.

COMMENTS

The approach we have taken to the overdetermined least squares problem is through the normal equations. There are other methods based on orthogonalization techniques. We will see the least squares problem again in *The QR Factorization* and *The Singular Value Decomposition*. The approach based on the normal equations is less accurate but relatively fast. The orthogonal approach is more accurate but usually slower. MATLAB's \ uses an orthogonal method. MATLAB has a function `leastsq` for solving general least squares problems.

Linear Transformations

ABSTRACT

We introduce the concept of linear transformation and discuss its relationships with matrix multiplication, bases, and similarity.

MATLAB COMMANDS

```
inv, rref, *, \
```

LINEAR ALGEBRA CONCEPTS

Linear Transformations, Coordinates, Change of Base, Base Representation, Projection

BACKGROUND

A function $T : \mathrm{R}^n \to \mathrm{R}^m$ is a **linear transformation** if

(1) $T(u + v) = T(u) + T(v)$

(2) $T(\alpha u) = \alpha T(u)$ for all scalars α

Examples of linear transformations are provided by matrix multiplication. If A is an $m \times n$ matrix, then we can define $T : \mathrm{R}^n \to \mathrm{R}^m$ by $T(x) = Ax$. The usual rules of matrix multiplication apply so that

(1) $T(u + v) = A(u + v) = Au + Av = T(u) + T(v)$

(2) $T(\alpha u) = A(\alpha u) = \alpha Au = \alpha T(u)$ for all scalars α

We conclude that this defines a linear transformation. The easiest way to check that a function is a linear transformation is to find a matrix A where $T(x) = Ax$. The next theorem shows that all linear transformations from R^n to R^m come from matrix multiplication.

Theorem. *If $T : \mathrm{R}^n \to \mathrm{R}^m$ is a linear transformation, then there is an $m \times n$ matrix A such that $T(x) = Ax$ for all $x \in \mathrm{R}^n$.*

PROOF: Recall the standard basis $e_1, \ldots, e_n$ and let $v_i = T(e_i)$ for $i = 1, \ldots, n$. Let $A = [v_1, \ldots, v_n]$. Now look at $x = [x_1, \ldots, x_n]^T$ Then $x = x_1 e_1 + \cdots + x_n e_n$ and

$$T(x) = T(x_1 e_1 + \cdots + x_n e_n) = x_1 T(e_1) + \cdots + x_n T(e_n)$$
$$= x_1 v_1 + \cdots + x_n v_n$$
$$= Ax$$

The last equation follows from Theorem 1 of *Subspaces*. ∎

We will now consider a variety of linear transformations. We will show that each of the functions defined here is a linear transformation by finding a matrix A such that $T(x) = Ax$.

Projection of a vector onto a vector

The projection of v onto w is $p = \frac{v \cdot w}{w \cdot w} w$. If we define $P = (\frac{1}{w^T * w}) w * w^T$, where we are assuming that w is an $n \times 1$ column vector (notice that $w * w^T$ is an $n \times n$ matrix, called the **outer product** see page 32) then

$$Pv = (\frac{1}{w^T * w}) w * w^T * v = \frac{w^T * v}{w^T * w} * v = p.$$

Thus projection is a linear transformation.

Projection of a vector perpendicular to a vector

The projection of v perpendicular to w is $q = v - p$. Define $Q = I_n - P$, where P is the matrix for projection. Since $Pv = p$, we see that

$$Qv = (I - P)v = Iv - Av = v - p = q$$

and thus perpendicular projection is a linear transformation.

Reflection of a vector across a vector

The reflection of v across w is $r = p - q$ where p is the projection and q is the perpendicular projection. The matrix for reflection is found by $r = p - q = p - (v - p) = -v + 2p = -Iv + 2Pv = (-I + 2P)v$, so $R = -I + 2P$ is the matrix for reflection and reflection is a linear transformation.

Reflection of a vector across the perpendicular of a vector

The reflection of v across the perpendicular to w is $s = q - p$ where p is the projection and q is the perpendicular projection. Since $s = q - p = -(p - q) = -(-I_n + 2P)v = (I_n - 2P)v$, the matrix for perpendicular reflection is

$$H = I_n - 2P = I_n - \frac{2}{w \cdot w} w * w^T$$

where w is an $n \times 1$ vector. This matrix is the Householder matrix from *Building Matrices* which will be used in *The QR Factorization*.

Projection of a vector into a subspace

The projection of a vector v onto a subspace W is described in *Projections*. If we represent W as the Column Space of a matrix A where rank$(A) = n$, then the formula $p = A(A^T A)^{-1} A^T v$ gives the projection, p, of v onto W. The matrix equation $P = A(A^T A)^{-1} A^T v$ shows that projection onto a subspace is a linear transformation.

Rotation

Recall from *Graphics* that rotation of a vector v through an angle θ is Rv where

$$R = \begin{bmatrix} \cos(\theta) & -\sin(\theta) \\ \sin(\theta) & \cos(\theta) \end{bmatrix}$$

Thus rotation is a linear transformation. To rotate an $n \times 1$ vector in the i, j−plane, use a Givens matrix (see page 'givens' of *Building Matrices*). The Givens matrices are used in *The QR Factorization*.

Scaling

To *scale* an $n \times 1$ vector in the ith coordinate, multiply the ith coordinate by α. We can view this as matrix multiplication by letting $E = $ ele2(n, α, i) where ele2 is an elementary matrix of type 2 from *Building Matrices* . Thus scaling is a linear transformation. To do an *expansion* or a *contraction*, scale all of the coordinates.

Shear

Shearing is achieved by multiplying a vector by an elementary matrix of type 3 (see page 33 of *Graphics*). Thus, shearing is a linear transformation.

The Dot Product

Let $a \in R^n$ and define $T : R^n \to R$ by $T(x) = a \cdot x$. We see that T is a linear transformation by taking A to be the row vector with entries $A(i) = a_i$. Then $T(x) = Ax$.

The Cross Product

Let $a \in R^3$ and define $T : R^3 \to R^3$ by $T(x) = a \times x$, the vector cross product. Let

$$A = \begin{bmatrix} 0 & -a_3 & a_2 \\ a_3 & 0 & -a_1 \\ -a_2 & a_1 & 0 \end{bmatrix}$$

Then $T(x) = Ax$. It is easily checked that $T(x) \cdot a = 0$ and $T(x) \cdot x = 0$.

Differentiation

Consider the function $D : P_n \to P_{n-1}$ on polynomials defined by $D(f(x)) = f'(x)$, the derivative of $f(x)$. The usual properties of differentiation, that is

$$f' + g' = f' + g' \qquad (\alpha f)' = \alpha f'$$

show that D is a linear transformation. To find a matrix for D consider

$$f(x) = a_1 x^{n-1} + a_2 x^{n-2} + \cdots + a_{n-1} x + a_n$$
$$D(f) = (n-1) a_1 x^{n-2} + (n-2) a_2 x^{n-3} + \cdots + a_{n-1}$$

Viewed as a transformation on vectors: $a = [a_1, a_2, \ldots, a_{n-1}, a_n]^T$

$$T(a) = [(n-1)a_1, (n-2)a_2, \ldots, a_{n-1}]^T$$

The $(n-1) \times n$ matrix which represents differentiation on polynomials of degree $n-1$ is

$$\begin{bmatrix} n-1 & 0 & \cdots & \cdots & 0 \\ 0 & n-2 & \ddots & & \vdots \\ \vdots & & \ddots & \ddots & \vdots \\ 0 & \cdots & 0 & 1 & 0 \end{bmatrix}$$

In MATLAB

```
D=[diag(3:-1:1),zeros(3,1)];
```

To differentiate the polynomial $f(x) = 2x^3 - x + 4$ we represent the polynomial as

```
f=[2 0 -1 4]';
D*f
```

This represents the polynomial $6x^2 - 1$.

Coordinates

If $u_1, \ldots, u_n$ is a basis and $x = \sum a_i u_i$, we call $a = [a_1, \ldots, a_n]$ the ***vector of coordinates of*** x ***with respect to*** $u_1, \ldots, u_n$ (see page 113.) Define $T : R^n \to R^n$ by

$$T(x) = \text{the vector of coordinates of } x \text{ with respect to } u_1, \ldots, u_n$$

If we let $U = [u_1, \ldots, u_n]$, then $x = \sum a_i u_i = Ua$ and $T(x) = U^{-1}x$.

Suppose that we view the columns of `magic(3)` as a basis for R^3 and we want to get the coordinates of $x = [1, 1, 1]^T$.

```
x=ones(3,1);
U=magic(3)
C=inv(U)
c=C*x
c(1)*U(:,1)+c(2)*U(:,2)+c(3)*U(:,3)
```

This will work for any vector.

```
x=rand(3,1); c=C*x
x-c(1)*U(:,1)+c(2)*U(:,2)+c(3)*U(:,3)
```

Change of Base

Now suppose that we have two bases $u_1, \ldots, u_n$ and $v_1, \ldots, v_n$. Given a vector a we can view a as the coordinates of another vector, namely, $w = \sum a(i)u_i$. We can find the vector of coordinates, b, of w with respect to $v_1, \ldots, v_n$, $w = \sum b(i)v_i$. The transformation $T(a) = b$ is the ***change of base transformation*** from coordinates for the u_i's to coordinates for the v_i's. The change of base transformation is a linear transformation. Let $U = [u_1, \ldots, u_n]$ and $V = [v_1, \ldots, v_n]$. We have the following relationships: $w = Ua$ and $Vb = w$. From this we get $T(a) = b = V^{-1}w = V^{-1}Ua$. In MATLAB this change of base matrix is `V\U`. It can also be found by row reducing, since `rref([V,U])=[eye(n),V`$^{-1}$`U]`. Returning to the previous example. Let `x,U,` and `c` be as given above. Now take a new basis.

```
V=compan(1:4);
C=V\U;
b=C*c
```

Recall that `c` is the vector of coordinates for `x` with respect to `U`. We will see that `b=C*c` is the vector of coordinates of `x` with respect to `V`.

```
x=b(1)*V(:,1)+b(2)*V(:,2)+b(3)*V(:,3)
```

Determining a linear transformation from a basis

Supposing that $u_1, \ldots, u_n$ is a basis for R^n and $v_1, \ldots, v_n$ are any vectors in R^m, there is a unique linear transformation $T : R^n \to R^m$ with the property that $T(u_i) = v_i$ for all $i = 1, \ldots, n$. The value $T(x)$ is defined by first finding the coordinates $a = [a_1, \ldots, a_n]$ for x with respect to $u_1, \ldots, u_n$, so that $x = \sum_i a_i u_i$, and then

$$T(x) = T(\sum_i a_i u_i) = \sum a_i T(u_i) = \sum_i a_i v_i$$

In matrix terms, if $U = [u_1, \ldots, u_n]$, $V = [v_1, \ldots, v_n]$ and $a = [a_1, \ldots, a_n]^T$, then we have the relationships $x = Ua$ and $T(x) = Va$. Thus $T(x) = Va = VU^{-1}x$. In MATLAB we choose a new basis for R^5 and define a transformation into R^3.

```
U=triu(magic(5))

V=rand(3,5)

A=V*inv(U)
```

We show that this does transform the first column of U into the first column of V. The other columns are left to you.

```
A*U(:,1)=V(:,1)
```

Base Representation of a Linear Transformation

Suppose that $T : \mathrm{R}^n \to \mathrm{R}^n$ is a linear transformation represented by $T(x) = Ax$ and $u_1, \ldots, u_n$ is a basis for R^n. Given a vector x we can get coordinates a for x with respect to $u_1, \ldots, u_n$ and coordinates b for $T(x)$ with respect to $u_1, \ldots, u_n$. The function $S : \mathrm{R}^n \to \mathrm{R}^n$ given by $S(a) = b$ is a linear transformation. To get a matrix B which represents S consider these equations: let $U = [u_1, \ldots, u_n]$

$$x = Ua, \qquad T(x) = Ax, \qquad T(x)' = Ub$$

Then

$$b = U^{-1}T(x) = U^{-1}Ax = U^{-1}AUa = Ba.$$

Notice that $B = U^{-1}AU$ is similar to A. See page 171 of *Eigenvalues* for more information on similarity.

PROBLEMS

1. Let U=rand(5) and V=pascal(5). Since U and V are invertible, the columns of U are a basis for R^5 and the columns of V are a basis for R^5. Find the change of base matrix C from the U–basis to the V–basis. Let x=(1:5)' find the coordinates a for x with respect to U and the coordinates b for x with respect to V. Check that C*a=b. Now find the change of base matrix D from the V–basis to the U-basis. Check that D*b=a.

2. Let U=eye(5), V=pascal(5) and suppose that $T : \mathrm{R}^5 \to \mathrm{R}^5$ has the property that T(U(:,i))=V(:,i) for $i = 1, \ldots, 5$. Find a matrix A where $T(x) = Ax$ for all $x \in \mathrm{R}^5$. Check your work by checking if A*U(:,i)=V(:,i).

3. Let U=compan(ones(6,1)), V=pascal(5). Suppose $T : \mathrm{R}^5 \to \mathrm{R}^5$, is a linear transformation with the property that T(U(:,i))=V(:,i) for i=1,...,5. Find a matrix A where $T(x) = Ax$ for all $x \in \mathrm{R}^5$. You can check your work by looking at A*U. What should the answer be? Now find a matrix B which represents T with respect to U. As a test for your work do the following: x=rand(5); find the coordinates a for x with respect to U; let b=Ba; check that b is the coordinate vector for Ax with respect to U. Show that A and B are similar by finding an invertible C where inv(C)*A*C=B. What is rank(A)? Can you find a matrix which represents T with respect to V?

4. You can graphically see the geometric transformations. Let

```
x=[0,1,1,0,0]
y =[0,0,1,1,0].
```
This will form a square. Now set the axis with
```
axis([-3,3,-3,3]
plot(x,y), hold
```
You can scale the x-axis with
```
S=eye(2); S(1,:)=2*S(1,:),
```
which is an elementary matrix of type 2. Now let
```
u=S*[x;y]; plot(u(1,:),u(2,:)).
```
You can shear with
```
A=eye(2); A(:,2)=2*A(:,1)+A(:,2);
```
which is an elementary matrix of type 3. Now let
```
v=A*[x;y]; plot(v(1,:),v(2,:)).
```
You can rotate with
```
R=[cos(pi/4), -sin(pi/4); sin(pi/4), cos(pi/4)],
w=R*[x;y]; plot(w(1,:),w(2,:));
```

5. When we successfully diagonalize a matrix A we find an invertible matrix P and a diagonal matrix D where $P^{-1}AP = D$. If we view A as a linear transformation using the standard basis, then we can view D as the matrix representation of that transformation in the basis given by the columns of P. Do this as follows: $D = PAP^{-1}$, so given a vector x, $c = P^{-1}x$ is the vector of coordinates of x with respect to P. $Dc = DP^{-1}x$ is the outcome of applying the transformation D, and $PDc = PDP^{-1}x = Ax$ converts back to standard coordiates. Using `A=magic(5)` and `x=ones(5,1)`, go through this computation. You can use `eig` from *Eigenvalues* to find P and D.

6. Consider the linear transformation $T : P_4 \to P_5$ defined by ***integration*** of the polynomial $f(x) = a_1x^3 + a_2x^2 + a_3x + a_4$ to $\frac{a_1}{4}x^4 + \frac{a_2}{3}x^3 + \frac{a_3}{2}x^2 + a_4x$. Find a matrix A which represents

$$T([a_1, a_2, \ldots, a_n]^T) = \left[\frac{a_1}{n}, \frac{a_2}{n-1}, \ldots, a_n, 0\right]^T$$

Try it for $n = 5$. What is $A * D$ where D is the differentiation matrix in problem 7.

7. Let `A=magic(6)`, `C=rand(6)`, and `B=inv(C)*A*C`. Compute `eig(A)`, `eig(B)`, `det(A)`, `det(B)`, `rank(A)`, and `rank(B)`.

8. Consider the following conic

$$3x^2 + 2xy + 3y^2 = 1$$

We want to get a graph of this conic. First notice that this can be represented by the matrix equation

$$[x \quad y]\begin{bmatrix} 3 & 1 \\ 1 & 3 \end{bmatrix}\begin{bmatrix} x \\ y \end{bmatrix} = 1$$

Let A be the coefficient matrix given above. Using `eig` find an invertible P and a diagonal D such that $P^{-1}AP = D$. Notice that MATLAB has returned an orthogonal matrix P, i.e.

$P^T = P^{-1}$. Let $[x1, y1] = [x, y]P$ so transposing both sides, $[x1, y1]^T = P^T[x, y]^T = P^{-1}[x, y]$. Thus we get

$$[\, x1 \quad y1 \,] D \begin{bmatrix} x1 \\ y1 \end{bmatrix} = 1$$

You can now plot this in $x1y1-$ coordinates. The easiest way to do this is to get plotting coordinates for the unit circle:

```
t=0:pi/20:2*pi; x=cos(t); y=sin(t);
```

The next step takes some justification. We have $[x, y] \cdot [x, y] = 1$ as the equation of the unit circle. Since the eigenvalues are both positive we can form $D^{-1/2}$ and let $[x2; y2] = D^{-1/2}[x; y]$. We note that $[x2, y2]D[x2; y2] = [x, y]D^{-1/2}DD^{-1/2}[x; y] = [x, y][x; y] = 1$. So now multiply

```
z=inv(sqrt(D))*[x;y]; x1=z(1,:); y1=z(2,:); plot(x1,y1)
```

Now you can get $xy-$ coordinates from the substitution

```
z=P*[x1;y1]; x=z(1,:); y=z(2,:); plot(x,y)
```

Notice how similar matrices and orthogonal matrices were used here. See *Symmetric Diagonalization* for more of this.

Error Correcting Codes

ABSTRACT

In transmitting data a main consideration is the the reliable reception of the data. If the data has been corrupted in any way, then it is necessary to either retransmit or attempt to correct the received message. In this project we set up some simple binary codes and schemes for detecting and correcting errors.

MATLAB COMMANDS

```
*, sum, min, floor, rref, '
```

LINEAR ALGEBRA CONCEPTS

Row Space, Null Space, Orthogonal Complement, Standard Basis, Dimension

BACKGROUND

Messages can be transmitted in a wide variety of ways. Noise can enter the path of transmission-reception through stray frequencies such as lightning or even a physical object coming between a laser transmitter and receiver. Inside a computer messages are sent from one chip to another and noise can be introduced through fluctuations in voltage. Even an error in software can make random changes in data which can also be regarded as noise. The purpose of error-correcting codes is to encode the message so that if noise does corrupt the message during transmission, it will be possible to detect and correct the message. One way to achieve this is to send the message more than once; redundant transmission. This method is necessary if the noise is of such a type that the entire message is obliterated. It is an expensive approach to the problem. If the noise only corrupts a small part of the message, then usually check-sums are used to correct the received message. It is this latter type of correction and detection that we will address here.

Throughout this project we will be using row vectors, rather than our usual convention of column vectors. Vector multiplication will place the vector to the left of the matrix. Instead of working with column spaces, we will work with row spaces.

A **message** is a binary vector of length k, that is, a vector with entries taken from $\{0, 1\}$. Until further notice we will use binary matrices. The operations of addition, subtraction, and multiplication will be done in mod 2 arithmetic. In this system $2 = 1 + 1 = 0$ and $1 = -1$, so that for all $a, b \in \{0, 1\}$, $a + a = 0$ and $a + b = a - b$. When we talk of subspaces, linear combinations and other concepts from linear algebra, it is understood that the scalars are in $\{0, 1\}$.

A **binary (n,k) code** where $k < n$ is given by a $k \times n$ matrix

$$C = [I_k, D]$$

where D is a $k \times (n - k)$ matrix. If v is a $1 \times k$ vector, then $vC = [v, vD]$ is a $1 \times n$ vector, called a **code word**. Notice that each code word consists of the message vector, v, and the $1 \times (n - k)$ vector vD. The set of all code words is a subspace, the Row Space of C. The matrix D is used for error detection and correction. The rank of C is k and the scalars come from $\{0, 1\}$, so, as there are 2^k possible messages, there are 2^k code words.

Consider the (5,3) code given by

$$C = \begin{bmatrix} 1 & 0 & 0 & 1 & 1 \\ 0 & 1 & 0 & 0 & 1 \\ 0 & 0 & 1 & 1 & 1 \end{bmatrix}.$$

In this code, a message is a binary vector of length 3, and a code word is an element of the Row Space of C.

The Row Space of C is called the **code of C** and C is called a **code generator.** The Row Space of C consists of all linear combinations

$$u = \alpha[1, 0, 0, 1, 1] + \beta[0, 1, 0, 0, 1] + \gamma[0, 0, 1, 1, 1] = [\alpha, \beta, \gamma]C$$

where $\alpha, \beta, \gamma \in \{0, 1\}$.

When we let α, β, and γ vary we get the the following vectors:

code word	weight
[0 0 0 0 0]	0
[0 0 1 1 1]	3
[0 1 0 0 1]	2
[0 1 1 1 0]	3
[1 0 0 1 1]	3
[1 0 1 0 0]	2
[1 1 0 1 0]	3
[1 1 1 0 1]	4

The **weight** of a binary vector, u, is the number of 1's which occur in u, *i.e.*

$$\text{weight}(u) = \sum_i u(i) = \text{sum}(u).$$

The weights of the code words of our example are listed. The **weight of the code** is

$$\min\{\text{weight}(u) : u \text{ is a code word and } u \neq \vec{0}\}.$$

Thus the example code has weight 2. At the end of this project is an m-file called enum.m. Type it into your directory and create the matrix C given above. To generate the code given above

```
E=(enum(3))'
code=E*C
code=mod(code,2)
```

Recall the function mod given on page 16 in the *MATLAB Tutorial.* You should get a complete listing of the code of C. The weights are easily computed

```
sum(code')'
```

To get the weight of the code, remove the row of zeros, and call

```
min(sum(code'))
```

The *parity check matrix* is the $(n - k) \times n$ matrix $P = [D^T, I_{n-k}]$ where D comes from the code generator C. In our example,

$$P = \begin{bmatrix} 1 & 0 & 1 & 1 & 0 \\ 1 & 1 & 1 & 0 & 1 \end{bmatrix}$$

is the parity check matrix. The Row Space of P is called the *dual code of C*. Using the parity check matrix there is an easy test to see if a vector is a code word. See *Projections* for the terminology in the next theorem.

Theorem 1. *The dual code of C is the orthogonal complement of the code of C. If P is the parity check matrix of C, then u is a code word if and only if $Pu^T = \vec{0}$*

PROOF: Let W=code of P and V=code of C. We first check that every row of P is perpendicular to every row of C. Let $v = [D(:, i)^T, e_i]$ be a row of P where $e_i = (0, \ldots, 1, \ldots, 0)$ with the 1 in the ith position, and let $w = [e_j, D(j, :)]$ be a row of C, then

$$v \cdot w = D(:, i) \cdot e_j + e_i \cdot D(j, :) = D(j, i) + D(j, i) = 0$$

where the arithmetic is done mod 2. Thus, $W \subseteq V^\perp$. Now the rank$(P) = \dim(W) = n - k$ and rank$(C) = k$, so that $\dim(V^\perp) = n - k$ and thus by Theorem 5 of *Dimension and Rank*, $W = V^\perp$. Now if P is the parity check matrix,

$$u \text{ is a code word} \iff u \text{ is in the Row Space of } C$$
$$\iff v \cdot u = 0 \text{ for all } v \text{ in the Row Space of } P$$
$$\iff Pu^T = \vec{0}$$

∎

The theorem is true even when we are not using binary matrices if we define the parity check matrix to be $P = [-D^T, I_{n-k}]$.

The equations $Pu^T = \vec{0}$ are called the *parity check equations.* If $u = vC = [v, vD]$ is the code word then the obvious decoding of u is v. If the received word has been corrupted badly enough we might get a code word and think that we are decoding correctly. Try this in MATLAB by entering the generating matrix C and the parity check matrix P given above and let

```
v=[1,1,1]
```

```
u=v*C
```

We can check that this is a code word (even though we know that it is) with the parity check matrix

```
P*u'
```

Now corrupt u in two places as follows:

```
w=[1,0,1,0,0]
```

```
P*w'
```

We see that w is also a code word. In fact

$$[1, 0, 1] * C$$

How would we know which message was sent, $[1, 1, 1]$ or $[1, 0, 1]$? This is a problem which we cannot resolve. However, if the received message is not a code word and we assume that the intended message is the unique code word which is closest to the received message, then we can decode. This is the method of nearest neighbor decoding.

Nearest Neighbor Decoding

Suppose that the binary vector u is received. **Nearest neighbor decoding** is described by nearngbr(u). We define nearngbr(u) = v if and only if

$$\text{weight}(u - v) = \min\{\text{weight}(u - w) : w \text{ is a code word}\}$$

This describes a well-defined function when there is a unique v with minimum weight($u - v$).

Now suppose in our example that $u = [1, 1, 1, 1, 1]$ is received. By checking $uP^T = [1, 0]$ we see that u is not a code word. Using nearest neighbor decoding we need to find the code word v where the weight of $u - v$ is minimal. But look here, $v = [1, 1, 1, 0, 1]$ is a code word and $u - v = [0, 0, 0, 1, 0]$ has weight 1 and $u - v$ has weight ≥ 2 for all other code words v, so nearngbr(u) = v. In the statement of the next theorem `floor` is the MATLAB function which satisfies $\texttt{floor}(x) = n \iff n \leq x < n + 1$.

Theorem 2. *If the weight of the code is d, then any received word with no more than $t = \texttt{floor}(\frac{d-1}{2})$ errors can be correctly decoded using nearest neighbor decoding.*

PROOF: Suppose that u is a received vector with t or fewer errors and that w is the code word sent, then

$$\text{weight}(w - u) \leq t.$$

Now suppose that v is the nearest neighbor. We must show that $v = w$. Since v is the nearest neighbor,

$$\text{weight}(v - u) \leq \text{weight}(w - u) \leq t.$$

Now if $v \neq w$, then since both w and v are in the Row Space so is $w - v$, and so $w - v$ is a nonzero code word. Thus

$$d \leq \text{weight}(w - v) \leq \text{weight}(w - u) + \text{weight}(u - v)$$
$$\leq 2t \leq d - 1 < d$$

We have a contradiction, thus $v = w$. ∎

Returning to our example, we have calculated the weight to be 2. By the Theorem 2, nearest neighbor decoding can correct $\texttt{floor}(\frac{2-1}{2}) = 0$ errors. Suppose that the received vector is $u = [1, 0, 1, 0, 1]$, then there are two nearest neighbors, $[1, 0, 1, 0, 0]$, and $u = [1, 1, 1, 0, 1]$. It follows that there is no unique nearest neighbor decoding of u and we cannot tell which message was sent.

Hamming Codes

The Hamming codes will detect and correct a single error and they do it efficiently. We will give a slightly backwards definition of a Hamming code in terms of the dual code. Let $n > 0$ be an integer and define $H(n)$ to be the $n \times (2^n - 1)$ matrix whose columns are the nonzero binary vectors of length n. For example,

$$H(3) = \begin{bmatrix} 0 & 0 & 0 & 1 & 1 & 1 & 1 \\ 0 & 1 & 1 & 0 & 0 & 1 & 1 \\ 1 & 0 & 1 & 0 & 1 & 0 & 1 \end{bmatrix}.$$

The **Hamming** $(2^n - 1, 2^n - n - 1)$ **Code** is the dual of the code generated by $H(n)$. More directly, $H(n)$ is a the parity check matrix for the Hamming code, that is, for all binary vectors v of length $2^n - 1$ and $H = H(n)$

$$u \text{ is a code word if and only if } Hu^T = \vec{0}$$

It is customary to arrange the columns of $H = H(n)$, as we have done with $H(3)$, so that $H(:, k)$ is the binary representation of k, i.e. for each $k = 1, \ldots, 2^n - 1$,

(1) $$k = H(n, k) + H(n - 1, k)2 + H(n - 2, k)2^2 + \ldots + H(1, k)2^{n-1}$$

To get a code generator for the Hamming code, let C be a matrix whose rows are a basis for the Null Space of H. Thus $HC^T = \vec{0}$, and so we can conclude that the Code of C = Row Space of C = Hamming Code. For those who have completed *The Null Space* and have the function $\texttt{nullbase}$ (see page 122,) take $H = H(3)$, the matrix given above and find a basis for the Null Space of H.

    C=nullbase(H)

By Theorem 1, the transposes of these basis vectors will be a basis for the code.

    C=C'

Now put the code generator into standard form using

    C=rref(C), C=mod(C,2)

Since H has $rank = n$, we see that the Null Space of H has dimension $2^n - n - 1$. Thus C is a $(2^n - n - 1) \times (2^n - 1)$ matrix.

Theorem 3. *For $n \geq 2$ the Hamming $(2^n - 1, 2^n - n - 1)$ code has weight 3.*

PROOF: I suggest that you use $H(3)$ given above for guidance through this proof. First we will show that there is a code word of weight 3. Recall $e_i = (0, \ldots, 0, 1, 0, \ldots, 0)$ where the 1 occurs in the ith position. In $H = H(n)$, the first three columns are

$$H(:, 1) = e_n, \ H(:, 2) = e_{n-1}, \ H(:, 3) = e_n + e_{n-1}.$$

Let $v = e_1 + e_2 + e_3$, then

$$vH^T = (1, 1, 1, 0, \ldots, 0)H^T$$
$$= (0, \ldots, 0, v \cdot H(n-1, :), v \cdot H(n, :))$$
$$= (0, \ldots, 0, 0, 0)$$

Thus, v is a code word and clearly v has weight 3.

We will now show that there are no code words of weight 1 or 2. Suppose that e_i is a code word, then $e_i H^T = \vec{0}$. Thus $H(i, :)^T = H(:, i) = \vec{0}$, which is not true.

Suppose the $e_i + e_j$ is a code word where $i \neq j$. Then $(e_i + e_j)H^T = \vec{0}$ and so $H(:, i) + H(:, j) = \vec{0}$, but mod 2, this means $H(:, i) = H(:, j)$ which is not true. ∎

We know that nearest neighbor decoding will correct a single error, but the beauty of Hamming codes is that the error correction is done directly, not by a search for the nearest neighbor.

Theorem 4. *Suppose that $H = H(n)$ for $n \geq 3$ and that y is a $1 \times (2^n - 1)$ vector which differs from a code word in only the kth bit, then*

$$b = yH^T \text{ is the binary representation of } k.$$

PROOF: Let $k = \sum_{j=1}^{n} b(j)2^{n-j}$, so that that b is the binary representation for k. Since $H(:, k)^T$ is also the binary representation for k, we see that $b = H(:, k)^T$ by the uniqueness of the binary representation. Now let $z = y + e_k$. This changes only the kth bit. Look at the parity check equations

$$zH^T = yH^T + e_k H^T = b + H(:, k)^T = \vec{0} \ (\text{mod } 2)$$

Thus z is a code word. Since the weight of the code is 3 by Theorem 3. We can correct one error using nearest neighbor decoding by Theorem 2. Since y is not a code word and weight$(y + z)$ = weight(e_k) = 1, z is the nearest neighbor to y and thus z is the unique code word which differs from y in a single bit. We conclude that k is the location of the error and z is the correctly decoded word. ∎

We used the nearest neighbor decoding method in the proof of this theorem, but we do not need to use it in the decoding method. We note that if there is more than one error, then the decoding $z = y + e_k$ will give us a code word, but it differs from the correct code word. In MATLAB make $H(3)$

```
H=enum(3); H=H(:,2:8);
```
Now form the code
```
C=nullbase(H); C=mod(H,2)';
```
Now create a received message.
```
y=[1 1 1 1 0 1 0];
b=mod(y*H',2)
```

We see from the parity check matrix that y is not a code word. By inspection of mod (y*H', 2) we see that the error occurs in y(2). Rather than finding the location by converting the base we can match y*H' with H

```
diff=mod(H-b'*ones(1,7),2)
```

Now we can see clearly that B matches the second column of H. To get the number of the column

```
[m,k]=min(sum(diff))
```

And change the value to decode.

```
y(k)=1-y(k)
```

This is a code word!

```
mod(y*h',2)
```

PROBLEMS

1. List all of the code words and determine the weight of the codes generated by the following code generators. How many errors can be corrected using nearest neighbor decoding?

$$A = \begin{bmatrix} 1 & 0 & 0 & 1 \\ 0 & 1 & 1 & 1 \end{bmatrix}$$

$$B = \begin{bmatrix} 1 & 0 & 1 & 1 & 0 \\ 0 & 1 & 0 & 1 & 1 \end{bmatrix}$$

$$C = \begin{bmatrix} 1 & 0 & 0 & 0 & 1 & 1 \\ 0 & 1 & 0 & 1 & 1 & 1 \\ 0 & 0 & 1 & 1 & 1 & 0 \end{bmatrix}$$

$$D = \begin{bmatrix} 1 & 0 & 0 & 1 & 1 & 0 \\ 0 & 1 & 0 & 1 & 0 & 1 \\ 0 & 0 & 1 & 0 & 1 & 1 \end{bmatrix}$$

$$E = \begin{bmatrix} 1 & 0 & 0 & 0 & 0 & 1 & 1 & 1 \\ 0 & 1 & 0 & 0 & 1 & 0 & 1 & 1 \\ 0 & 0 & 1 & 0 & 1 & 1 & 0 & 1 \\ 0 & 0 & 0 & 1 & 1 & 1 & 1 & 0 \end{bmatrix}$$

2. For each of the codes in problem 1 find a vector which is has two nearest neighbors and hence cannot be unambiguously decoded by nearest neighbor decoding.

3. Write a MATLAB function w=nearngbr(C,v) which returns the code word which is the nearest neighbor of the vector v in the code generated by the matrix C. Hint: Use an outer product (see page 32 of *Building Matrices*) to get a matrix which has the same size as C and every row is just the vector v. Type help max.

4. Find the parity check matrix for each of the code generators in problem 1. If code is the listing of the code of one of these code generators and P is the parity check matrix, what is P*code'?

5. Write a MATLAB function H=ham(n) which finds the $n \times (2^n - 1)$ parity check matrix for the Hamming $(2^n - 1, 2^n - n - 1)$ code. Hint: use enum.

6. Using the method described in the Background discussion, find a code generator C for the Hamming $(15,11)$ code. Look closely at the last row of the matrix C. Notice how C does not quite have the form $[I_{11}, D]$. This is not a serious problem since the code is the Row Space of C, but it does raise the question of how to decode back to the original message. Recall that if $C = [I_k, D]$, then the code word is $vC = [v, vD]$, and after error correction, we can decode back to the original message by taking the first k columns of vC. With the code generator we have obtained for the $(15,11)$ code we cannot do this. We can decode by taking the first 10 columns and the 12th column. More generally, we do this by finding the columns which correspond to the leading variables.

7. Write a MATLAB function [C,L]=hamgen(n) which finds a code generator C for the Hamming $(2^n - 1, 2^n - n - 1)$ code. C should be in reduced row echelon form. L is a binary vector obtained from the function lead (see page 27 of *Systems of Linear Equations*) which will allow us to strip off the parity check bits when decoding. Try your hamgen on the Hamming $(7, 4)$ and $(15, 11)$ codes.

8. Write a MATLAB function z=hamdcode(n,y) which takes as input a $1 \times (2^n - 1)$ vector y, to be thought of as the received word, and returns a $1 \times (2^n - 1)$ vector which has been decoded using the method of Theorem 4. Try your hamdcode on the n=4, v=zeros(15,1), v([3, 5, 11,4])=ones(1,5).

9. Write a MATLAB function P=parity(C) which is the parity check matrix for the code given by C. We are going to create a noisy transmission channel. Define the function

```
function N=noise(n,tol)
   N = (rand(1,n) < tol);
```

This creates a random $1 \times n$ binary vector, N, and if $0 \leq tol \leq 1$, then tol*n is the expected number of 1's. We can regard this as random noise in a channel. If a binary vector v is transmitted, then v+N is the corrupted vector. Suppose that we can send messages of either length 4 or length 11 in a channel which has noise determined by tol=.1. Would you use a Hamming (7,4) code or a Hamming (15,11) code for reliable error detection and correction? Run 100 trials using each code, keeping track of the number of times that the decoded message differs from the original message sent. You can randomly generate binary messages in the same way that we have generated the noise, in fact, if you want, for this you can randomly generate tol.

10. A message of 44 bits is to be sent. If we use a (7,4) code we can send 11 4-bit "packets." If we use a (15,11) code we can use 4 11-bit packets. How many flops (see *Flops*) are used in sending and decoding the message with the (7,4) code? How many with the (15,11) code? You should assume that you already have the code generator matrix [C,L]=hamgen(n) and the parity check matrix H so that their construction is not counted in the flop count.

COMMENTS

The following function is useful for generating Hamming codes.

```
function B=enum(n)

% This gives an n by 2^n matrix
% whose columns constituted all
% n x 1 binary vectors.

if n==1, B=[0,1];
else
   B=[zeros(n,2^(n-1)),ones(n,2^(n-1))];
   C=enum(n-1);
   B(2:n,:)=[C,C];
end
```

Splines

ABSTRACT

A cubic spline is a piecewise polynomial function where each of the polynomial pieces is a cubic polynomial. A cubic spline preserves continuity, first derivatives and second derivatives at the interior knots. Hermite cubic splines allow the user to specify the derivative at the knots.

MATLAB COMMANDS

    rref, plot, mkpp, ppval, unmkpp, spline, interp1, polyplot

LINEAR ALGEBRA CONCEPTS

Polynomials, Calculus, Systems of Linear Equations, Lagrange Interpolation

BACKGROUND

We begin by considering points $(x_1, y_1), \ldots, (x_{n+1}, y_{n+1})$, where $x_1 < x_2 < \ldots < x_{n+1}$, which are to be interpolation points or **knots**. A **piecewise polynomial function** is a function $S : [x_1, x_{n+1}] \to \mathrm{R}$ built from polynomials, $p_1(x), \ldots, p_n(x)$ such that

$$
S(x) = \begin{cases}
p_1(x) & \text{if } x_1 \leq x \leq x_2 \\
p_2(x) & \text{if } x_2 \leq x \leq x_3 \\
\vdots & \vdots \\
p_n(x) & \text{if } x_n \leq x \leq x_{n+1}
\end{cases}
$$

If we want the function S to **interpolate** data, it will satisfy the interpolation conditions

$$
p_i(x_i) = y_i \quad \text{for } i = 2, \ldots, n
$$

and

$$
p_i(x_{i+1}) = y_{i+1} \quad \text{for } i = 1, \ldots, n - 1
$$

These conditions are most easily met by making lines between the points (x_i, y_i) and (x_{i+1}, y_{i+1}). This is called a **line spline** (see page 132 of *Least Squares*.) This is easy to do using the point-slope formula for a line:

$$
p_i(x) = \frac{y_{i+1} - y_i}{x_{i+1} - x_i}(x - x_i) + y_i
$$

We will take a linear algebra approach. Let

$$
p_i(x) = a_i + b_i(x - x_i)
$$

for $i = 1, \ldots, n$. Looking at the interpolation conditions we get

$$
p_i(x_i) = y_i
$$

which immediately gives $a_i = y_i$ and

$$p_i(x_{i+1}) = y_{i+1}$$

which gives

$$y_2 = b_1(x_2 - x_1) + y_1$$
$$y_3 = b_2(x_3 - x_2) + y_2$$
$$\vdots$$
$$y_{n+1} = b_n(x_{n+1} - x_n) + y_n$$

This produces the matrix equation

$$\begin{bmatrix} x_2 - x_1 & 0 & \cdots & 0 \\ 0 & x_3 - x_2 & & 0 \\ \vdots & & \ddots & \vdots \\ 0 & 0 & \cdots & x_{n+1} - x_n \end{bmatrix} \begin{bmatrix} b_1 \\ b_2 \\ \vdots \\ b_n \end{bmatrix} = \begin{bmatrix} y_2 - y1 \\ y_3 - y2 \\ \vdots \\ y_{n+1} - y_n \end{bmatrix}$$

Notice that we get a unique solution. Altogether there are six unknowns, $a_1, a_2, a_3, b_1, b_2, b_3$ and there are six equations obtained from the interpolation conditions. Try this in MATLAB with the points (0,2), (1,-1), (2.5,3), (3,0).

```
x=[0; 1; 2.5; 3];
y=[2; -1; 3; 0];
X=diff(x); Y=diff(y);
A=diag(X);
b=A\Y;
```

We have used the MATLAB function `diff`. This function computes the differences.

```
diff(x)=[x(2)-x(1), x(3)-x(2),...,x(n)-x(n-1)]
```

Now put this information into a single matrix.

```
C=[b,y(1:3)]'
pp=mkpp(x,C);
xx=0:.01:3;
yy=ppval(pp,xx);
plot(xx,yy,x,y,'o')
```

In this last group we used some of the piecewise polynomial functions of MATLAB in the evaluation of our line spline, namely `mkpp` and `ppval`. There is a special "pp" data structure that MATLAB uses for splines. The function `mkpp` creates this data structure from the coefficient matrix, C, for the spline and the break points, x, between the polynomials. The function `ppval` evaluates the spline at the input, xx, using the pp data structure. Notice that we used a coefficient matrix C which has the following form:

$$C = \begin{bmatrix} b_1 & a_1 \\ b_2 & a_2 \\ b_3 & a_3 \end{bmatrix}.$$

In general we will represent the coefficients of the spline as

$$C = \begin{bmatrix} d_1 & c_1 & b_1 & a_1 \\ d_2 & c_2 & b_2 & a_2 \\ \vdots & \vdots & \vdots & \vdots \\ d_n & c_n & b_n & a_n \end{bmatrix}$$

The MATLAB function `interp1` will find the output of the line spline directly.

```
yy=interp1(x,y,xx);
plot(xx,yy,x,y,'o')
```

To get a better looking curve through this data, we will go to cubic polynomials. The reason for going to cubics is that we will be able to control not only the continuity (with the interpolation conditions) but also the derivatives and the concavity. Consider now cubic polynomials

$$p_i(x) = a_i + b_i(x - x_i) + c_i(x - x_i)^2 + d_i(x - x_i)^3 \quad \text{for } i = 1, \ldots, n$$

There are $4n$ coefficients to be determined for $p_1, \ldots, p_n$. The interpolation conditions only provide $2n$ equations. Thus there are $2n$ free variables in the solutions to the linear system. For each polynomial we get the two equations:

$$y_i = a_i + b_i(x_i - x_i) + c_i(x_i - x_i)^2 + d_i(x_i - x_i)^3$$
$$y_{i+1} = a_i + b_i(x_{i+1} - x_i) + c_i(x_{i+1} - x_i)^2 + d_i(x_{i+1} - x_i)^3$$

These give the matrix equation

$$\begin{bmatrix} 1 & 0 & 0 & 0 \\ 1 & x_{i+1} - x_i & (x_{i+1} - x_i)^2 & (x_{i+1} - x_i)^3 \end{bmatrix} \begin{bmatrix} a_i \\ b_i \\ c_i \\ d_i \end{bmatrix} = \begin{bmatrix} y_i \\ y_{i+1} \end{bmatrix}$$

In MATLAB we can see some of the possibilities.

```
x=[0; 1; 2.5; 3]; y=[2; -1; 3; 0];
A=[1, 0, 0, 0; 1, x(2)-x(1), (x(2)-x(1))^2, (x(2)-x(1))^3]
R=rref([A,y(1:2)])
```

From this we get several possibilities for $p_1(x)$. We will be using `polyplot` from page 42 of *Graphics*. For example

```
C=zeros(4,1);
B=R(:,1:4); b=R(:,5);
C(1:2)=b-B*C;
clf
plot(x,y,'*')
polyplot(0,1,C(4:-1:1)')
```

Well, that just gave us the line back. Try some other values for the free variables.

```
C=zeros(4,1);
C(3:4)=[1;1];
C(1:2)=b-B*C;
polyplot(0,1,C(4:-1:1)')
```

Hermite Cubic Interpolation

One way to take the ambiguity out of cubic interpolation is to specify the derivatives at each of the knots. Thus in addition to the knots $(x_1, y_1), \ldots, (x_{n+1}, y_{n+1})$ also specify the derivative values

$$(x_1, y_1'), \ldots, (x_{n+1}, y_{n+1}').$$

To this add the derivative conditions:

$$p_i'(x_i) = y_i'$$
$$p_i'(x_{i+1}) = y_{i+1}'$$

This will provide $2n$ extra conditions, which should be enough to determine the cubic pieces. This type of interpolation is called **Hermite Cubic Interpolation** and the resulting function $S(x)$ is called a **Hermite cubic spline.** The next theorem tells us that the spline is determined by these conditions.

Theorem 1. *Given points (x_1, y_1) and (x_2, y_2) and derivative values y_1' and y_2' there is a unique cubic polynomial*

$$p(x) = dx^3 + cx^2 + bx + a$$

such that

$$p(x_1) = y_1 \quad and \quad p(x_2) = y_2$$
$$p'(x_1) = y_1' \quad and \quad p'(x_2) = y_2'$$

PROOF: Substituting x_1 and x_2 we get

$$y_1 = p(x_1) = dx_1^3 + cx_1^2 + bx_1 + a_1$$
$$y_2 = p(x_2) = dx_2^3 + cx_2^2 + b_1 x_2 + a_1$$
$$y_1' = p'(x_1) = 3dx_1^2 + 2cx_1 + b_1$$
$$y_2' = p'(x_2) = 3dx_2^2 + 2cx_2 + b_1$$

Which yields the matrix equation

$$\begin{bmatrix} x_1^3 & x_1^2 & x_1 & 1 \\ x_2^3 & x_2^2 & x_2 & 1 \\ 3x_1^2 & 2x_1 & 1 & 0 \\ 3x_2^2 & 2x_2 & 1 & 0 \end{bmatrix} \begin{bmatrix} d \\ c \\ b \\ a \end{bmatrix} = \begin{bmatrix} y_1 \\ y_2 \\ y_1' \\ y_2' \end{bmatrix}$$

It is easy to check that the coefficient matrix is invertible if $x_1 \neq x_2$ ∎

Let us return to writing the polynomials in shifted form:

$$p_1(x) = a_1 + b_1(x - x_1) + c_1(x - x_1)^2 + d_1(x - x_1)^3$$

The matrix equation in the theorem becomes

$$
\begin{bmatrix}
0 & 0 & 0 & 1 \\
(x_2 - x_1)^3 & (x_2 - x_1)^2 & x_2 - x_1 & 1 \\
0 & 0 & 1 & 0 \\
3(x_2 - x_1)^2 & 2(x_2 - x_1) & 1 & 0
\end{bmatrix}
\begin{bmatrix}
d \\ c \\ b \\ a
\end{bmatrix}
=
\begin{bmatrix}
y_1 \\ y_2 \\ y_1' \\ y_2'
\end{bmatrix}
$$

In MATLAB specify the derivatives at the knots: $(0, 2), (1, -1), (2.5, 3), (3, 0)$ to be $.5, .2, -5, 0$.

```
x=[0; 1; 2.5; 3]; y=[2; -1; 3; 0]; der=[.5; .2; -5; 0];
dif=x(2)-x(1);
A=[0, 0, 0, 1; dif^3,dif^2, dif, 1;
0, 0, 1, 0; 3*dif^2,2*dif, 1, 0];
Y=[y(1:2);der(1:2)];
s=A\Y; C=s';
pp=mkpp([0,1],C);
xx=0:.01:1;
yy=ppval(pp,xx);
plot(xx,yy)
```

So it looks like we are off to a good start. Repeat this process for the other knots.

```
dif=x(3)-x(2);
A=[0, 0, 0, 1; dif^3,dif^2, dif, 1;
0, 0, 1, 0; 3*dif^2,2*dif, 1, 0];
Y=[y(2:3);der(2:3)];
s=A\Y; C=[C;s'];
dif=x(4)-x(3);
A=[0, 0, 0, 1; dif^3,dif^2, dif, 1;
0, 0, 1, 0; 3*dif^2,2*dif, 1, 0];
Y=[y(3:4);der(3:4)];
s=A\Y; C=[C;s'];
```

We should now have the coefficient matrix. Apply the pp functions to get a plot.

```
pp=mkpp(x,C);
xx=0:.01:3;
yy=ppval(pp,xx);
plot(xx,yy,x,y'*')
```

It would be interesting to see what the derivatives at the knots look like. We can get plots using the utility vecplot from page 43 in *Graphics*.

```
hold
u=[x';y']
```

```
v=[x'+1;der'+y']
vecplot(u(:,1),v(:,1))
vecplot(u(:,2),v(:,2))
vecplot(u(:,3),v(:,3))
vecplot(u(:,4),v(:,4))
```

Cubic Splines

Hermite Interpolation makes sense if you have reasonable data about the derivatives. In the absence of that data it is still possible to find a spline that interpolates the data. This is done by requiring, in addition to the interpolation conditions, that the first and second derivatives be equal at the interior knot points. That is,

$$
\begin{aligned}
p_i(x_i) &= y_i & \text{for } i = 1, \ldots, n \\
p_i(x_{i+1}) &= y_{i+1} & \text{for } i = 1, \ldots, n \\
p'_i(x_{i+1}) &= p'_{i+1}(x_{i+1}) & \text{for } i = 1, \ldots, n - 1 \\
p''_i(x_{i+1}) &= p''_{i+1}(x_{i+1}) & \text{for } i = 1, \ldots, n - 1.
\end{aligned}
$$

Counting the equations here, we have $2n$ equations from the interpolation conditions, $n - 1$ equations from the first derivatives, and $n - 1$ equations from the second derivatives, for a total of $4n - 2$ equations.

There are a number of ways to determine the system by adding additional equations. The ***natural spline condition*** requires

$$
p''_1(x_1) = 0 \quad \text{and} \quad p''_n(x_{n+1}) = 0.
$$

These conditions are rather arbitrary and may not reflect the underlying behavior we are trying to approximate. ***Clamped splines*** allow the user to supply the first derivatives d_1 and d_2 at the end points giving the two additional equations

$$
p'_1(x_1) = d_1
$$
$$
p'_n(x_{n+1}) = d_2
$$

This requires that the user have some additional information about the function. Another way to fill in the missing equations is the ***not-a-knot condition*** which requires that the third derivative $S'''(x)$ be continuous at x_2 and x_n.

The MATLAB function `spline` is an implementation of the not-a-knot condition. Returning to our data set:

```
x=[0; 1; 2.5; 3]; y=[2; -1; 3; 0];
xx=0:.01:3;
yy=spline(x,y,xx);
plot(xx,yy,x,y,'o')
```

PROBLEMS

The first 5 problems are to be completed using the MATLAB function `spline`.

1. Plot a not-a-knot cubic spline through the points determined by

    x=0:9

    y=[-2,1,0,2,1,-2,0,-2,1,1]

2. Plot a not-a-knot spline which approximates $\cos(x)$ on the interval $[-\pi, \pi]$ using 10 equally spaced knots. Plot $\cos(x)$ over the spline to see how well the spline approximates it.

3. Plot a not-a-knot spline which approximates the Runge function $\frac{1}{1+25x^2}$ on the interval $[-1, 1]$ using 10 regularly spaced knots. Now try this using 10 Chebyshev knots (see page 57 of *Lagrange Interpolation* for the definition of the Chebyshev points). Plot the two splines together with the function.

4. Plot a not-knot-function which approximates `abs(x)` on the interval $[-1, 1]$ using first 10 regularly spaced knots and then 10 Chebyshev knots.

5. Plot a not-a-knot function which approximates `ceil(x)` on the interval $[-1, 2]$ using 10 regularly spaced knots followed by 10 Chebyshev knots.

6. Write a MATLAB function `C=hermite(d,der)` which finds the $n - 1 \times 4$ matrix of coefficients for the cubic polynomials as given above. d is a $2 \times n$ matrix of points, where a point (x_i, y_i) is given by x_i =d(1, i) and y_i =d(2, i). We assume that `d(1,1)` < `d(1,2)` < $\cdots$ < `d(1,n)`. der is the $1 \times n$ matrix of derivative values where y_i' =der(i).

7. Test your `hermite` by checking that it will recover the cubic polynomial $2x^3 + x + 3$ on $[1, 2]$ with the data

$$d = \begin{bmatrix} 1 & 2 \\ 6 & 21 \end{bmatrix}$$

The derivative is $6x^2 + 1$ so that der=[7, 25].

8. Write a MATLAB function `hermplot(C,d)` which plots the spline given by coefficients from C after evaluating at 20 points between each data point.

9. Plot the Runge function $f(x) = \frac{1}{25x^2+1}$ on the interval $[-3, 3]$. Now use a Hermite Cubic Spline with

$$d = \begin{bmatrix} -3 & -2 & -1 & 0 & 1 & 2 & 3 \\ f(-3) & f(-2) & f(-1) & f(0) & f(1) & f(2) & f(3) \end{bmatrix}$$

and der=zeros(1,7). Use `hermplot` to plot and compare the results. Now try

    der=[1,1,1,0,-1,-1,-1]

The derivative of the Runge function is $f'(x) = \frac{-50x}{(25x^2+1)^2}$. Try the derivative values at the data points for der. Which values for der give the best approximation?

10. Try problem 4 using the Chebyshev points as in problem 3.

11. Using the following data

$$d = \begin{bmatrix} 0 & 1 & 2 & 3 & 4 \\ 0 & 1 & 0 & 1 & 0 \end{bmatrix}$$

Find values for *der* which give a visually pleasing spline.

12. We are going to use a Hermite Cubic to approximate the value of log(2). The definition of $\log(x)$ is

$$\log(x) = \int_1^x 1/t \ dt$$

Get a cubic spline approximation to $1/x$ using `spline` with the data

```
x=[1, 3/2, 2];
y=[1, 2/3, 1/2];
```

Thus log(2) will be approximated by

$$\int_1^2 S(x) \ dx$$

With the two cubic polynomials $p_1(x)$ and $p_2(x)$ which make up $S(x)$ it is easy to integrate p_1 and p_2 and evaluate them to get your approximation to log(2).

15. Use the technique described in problem 12 to find a value for

$$\int_0^\pi \frac{\sin(x)}{x} \ dx$$

Recall that while this function is undefined at $x = 0$, $\lim_{x \to 0} \frac{\sin(x)}{x} = 1$

Orthonormal Bases

ABSTRACT

The notions of an orthogonal set of vectors and an orthonormal basis are introduced. Some applications to projections and solving equations are given. Finally, we give the classical Gram-Schmidt method for finding an orthonormal basis.

MATLAB COMMANDS
    orth

LINEAR ALGEBRA CONCEPTS
Orthogonal, Orthonormal, Gram-Schmidt Process

BACKGROUND

Vectors $w_1, \ldots, w_n \in \mathbf{R}^m$, where each $w_i \neq \vec{0}$, are **orthogonal** if $w_i \cdot w_j = 0$ for $i \neq j$. If we also require that $w_i \cdot w_i = 1$, then we say they are **orthonormal**. Since $\|w\|^2 = w \cdot w$, this latter condition says that the magnitude of each vector is 1. A vector w where $\|w\| = 1$ is called a **unit vector.** The difference between orthogonal and orthonormal is minor since we can **normalize** w into a unit vector u as follows:

$$u = \frac{1}{\|w\|} w.$$

If A is a matrix whose columns are orthonormal and $A = [w_1, \ldots, w_n]$, then the (i, j) entry of $A^T A$ is $w_i \cdot w_j$, which is 0 if $i \neq j$ and 1 if $i = j$. Thus

$$A^T A = I_n.$$

A is called an **orthogonal matrix** if A is square and $A^T = A^{-1}$. (Watch the terminology! An *orthogonal matrix* has *orthonormal* columns.) Unless an orthogonal matrix A is square AA^T will not be the identity. The condition, $A^T A = I_n$, is equivalent to the columns of A being orthonormal. MATLAB has a function which creates matrices with orthonormal columns:

    A=rand(5,3)
    A=orth(A)
    A'*A
    A*A'

Theorem 1. *Let $A = [w_1, \ldots, w_n]$ and suppose that the columns of A are orthogonal. Then the solution to $Ax = b$ is given by $x_i = \frac{w_i \cdot b}{w_i \cdot w_i}$. If the columns of A are orthonormal, then the solutions are given by $x_i = w_i \cdot b$*

The reader who is familiar with projections (see *Projections*) will recognize x_i from the projection formula. To solve $Ax = b$ where A is orthogonal is actually much easier than this since $A^{-1} = A^T$.

```
A=orth(rand(5))
b=rand(5,1)
x=A'*b
```

Check this with

```
A*x-b
```

Theorem 2. *If $w_1, \ldots, w_n$ are orthogonal, then they are linearly independent.*

The next result gives a simplified projection formula for the case of an orthogonal basis.

Theorem 3. *Let $w_1, \ldots, w_n$ be orthogonal column vectors and $A = [w_1, \ldots, w_n]$. Then*

$$proj(A, b) = \sum_{i=1}^{n} \frac{w_i \cdot b}{w_i \cdot w_i} w_i.$$

If $w_1, \ldots, w_n$ are orthonormal, then $w = \sum_{i=1}^{n}(w_i \cdot b)w_i$.

The projection matrix is simplified in the presence of a matrix A with orthonormal columns. Notice that the projection of b onto the Column Space of A is

$$w = A(A^T A)^{-1} A^T b = A A^T b.$$

Thus when the columns or A are orthonormal, the mysterious $A A^T$ is seen to be the projection matrix. In MATLAB

```
A=orth(rand(5,3))
P=A*A'
b=ones(5,1)
w=P*b
```

This gives us the projection, to check

```
(b-w)'*w
```

The Gram-Schmidt Process

We now consider the problem of finding an orthonormal basis for the Column Space of A. Suppose that the linearly independent vectors $v_1, \ldots, v_n$ are the columns of A. The *standard Gram-Schmidt Orthogonalization Process* finds vectors $w_1, \ldots, w_n$ by

$$w_1 = v_1$$

and for $k = 2, \ldots, n$

$$w_k = v_k - proj(B_{k-1}, v_k)$$

where $B_{k-1} = [w_1, \ldots, w_{k-1}]$. Thus w_k is the projection of v_k perpendicular to the Column Space of B_{k-1}. If you normalize w_k immediately after the assignment, then the columns of B_{k-1} will be orthonormal, and thus the projection formula will simplify to $\text{proj}(B_{k-1}, v) = B_{k-1}B_{k-1}^T v$.

Theorem 4. *The vectors produced by the Gram-Schmidt process are an orthonormal basis for the Column Space of A*

We can illustrate this in MATLAB. First choose a random set of linearly independent vectors.

```
V=rand(5,3)
v1=V(:,1)
v1=(1/norm(v1))*v1
B=[v1]
v2=V(:,2)-B*B'*V(:,2)
v2=(1/norm(v2))*v2
B=[B,v2]
v3=V(:,3)-B*B'*V(:,3)
v3=(1/norm(v3))*v3
B=[B,v3]
```

We can check orthonormality with

```
B'*B
```

We can check that A and B have the same column spaces with

```
rank(V)==rank([V,B])
```

The Modified Gram-Schmidt Process

You will notice that in the Gram-Schmidt Process we assign $w_k = v_k - \text{proj}(B_{k-1}, v_k)$. If v_k is close to the Column Space of B_{k-1}, then we will be subtracting nearly equal vectors. Subtracting nearly equal numbers is a source of round-off error (see page 67 of *Errors.*) As a result the Gram-Schmidt Process is unstable (see problem 4.) While v_k may be close to the Column Space of B_{k-1} it is not necessarily close to any of the vectors $w_1, \ldots, w_{k-1}$ which generate the space. Thus a simple alteration to the Gram-Schmidt Process helps to correct this flaw. The result is the ***Modified Gram Schmidt Process.***

In the Gram-Schmidt process at the kth iteration we assign

$$w_k = v_k - \text{proj}(B_{k-1}, v_k).$$

In the Modified Gram-Schmidt process

At the kth iteration, reassign the vector v_j

$$v_j = v_j - \text{proj}(w_{k-1}, v_j)$$

for each $j = k, \ldots, n$.

The new vector is the projection of v_j perpendicular to w_{k-1}. Finally define $w_k = v_k/\|v_k\|$. We can do this in MATLAB

```
V=rand(5,3)
V(:,1)=(1/norm(V(:,1)))*V(:,1)
```

Now we project the remaining two vectors perpendicular to V(:,1).

```
V(:,2)=V(:,2)-V(:,2)'*V(:,1)*V(:,1)
V(:,3)=V(:,3)-V(:,3)'*V(:,1)*V(:,1)
```

We normalize V(:,2)

```
V(:,2)=(1/norm(V(:,2)))*V(:,2)
```

and project the remaining vector perpendicular to V(:,2)

```
V(:,3)=V(:,3)-V(:,3)'*V(:,2)*V(:,2)
V(:,3)=(1/norm(V(:,3)))*V(:,3)
```

Check orthogonality with

```
V'*V
```

Those readers familiar with the norm concept (see *Norms and Condition Numbers*) can appreciate some results which relate the 2-norm to orthogonal matrices.

Theorem 5. *Suppose that A is orthogonal, then*

    (1) $x \cdot y = Ax \cdot Ay$
    (2) $x \cdot y = 0$ *if and only if* $Ax \cdot Ay = 0$
    (3) $\|Ax\|_2 = \|x\|_2$
    (4) $\|A\|_2 = 1$ *and so* $\kappa_2(A) = 1$.
    (5) $\|AB\|_2 = \|B\|_2 = \|BA\|_2$

PROOF: (1) $Ax \cdot Ay = x^T A^T Ay = x^T y = x \cdot y$.
(2) Immediate from (1).
(3) $\|Ax\|_2^2 = Ax \cdot Ax = x \cdot x = \|x\|_2^2$.
(4) $\|A\|_2 = \max_{\|x\|=1} \|Ax\|_2 = 1$, by (3).
(5) $\|AB\|_2 \leq \|A\|_2 \cdot \|B\|_2 = \|B\|_2$ and $\|B\|_2 = \|A^T AB\|_2 \leq \|A^T\|_2 \cdot \|AB\|_2 = \|AB\|_2$.
Similarly, for BA, since A^T is orthogonal.

∎

PROBLEMS

1. You can generate some matrices with orthonormal columns using MATLAB's orth. Try it on these matrices rand(7,4), magic(8), list(5), and hilb(8). In each case multiply A'*A to check for orthogonality. The matrix list is from page 34.

2. Write a MATLAB function B=grmsch(A) which uses the standard Gram Schmidt Process to find an orthonormal basis for the Column Space of A.

3. Test your function on rand(5), compan(ones(1,5)) and hilb(6). Multiply A'*A to check for orthogonality.

4. The following is a modification of an example due to Noble and Daniel. Let

```
A=eps*eye(5)+ones(5).
```

Run B=grmsch(A) and test for its accuracy by multiplying B'*B. The entries of B'*B are $w \cdot u$ where w and u are columns of B. We know from the formula $w \cdot u = \|w\|\|u\| \cos(\theta)$ and the fact that the columns have magnitude 1 that acos(B'*B) gives the angles between the columns of B. Using the formula *degrees*=$(\frac{180}{\pi})$ *radians,* you can express the angles between the columns in degrees. Are you satisfied with the outcome? What happens when you use MATLAB's orth? Is this consistent with MATLAB's rank(A)?

5. Write a MATLAB function B=modgram(A) which uses the Modified Gram Schmidt Process to find an orthonormal basis for the Column Space of A.

6. Test your modgram on the matrices in problem 3. Check the accuracy of modgram on these matrices. Now try modgram on the matrix in problem 4. How does the output compare with gram?

7. The Householder matrix $H = I_n - (\frac{2}{v^T v})vv^T$ where v is an $n \times 1$ column vector is defined in *Building Matrices*. (see page 33) Let v=rand(8,1) and form the Householder matrix H for v. Show that this H is orthogonal. See *The QR Factorization* for a proof that H is orthogonal for all choices of v.

8. The Givens matrix $G = $ givrot(n, i, j, c, s) where $c^2 + s^2 = 1$ is defined in *Building Matrices* (see page 'givens'.) Let t=rand, s=sin(t), and c=cos(t), and form the Givens matrix G=givrot(8,3,7,c,s). Show that G is orthogonal.

9. A QR factorization of a matrix A writes $A = QR$ where Q is orthogonal and R is upper triangular. Let A=rand(5) and apply MATLAB's qr to get a QR factorization of A. How is the 2-norm of A related to the 2-norm of R? Can you prove this? Apply qr to the matrix in problem 4. Now check Q for accuracy by computing the angles between the columns of Q. What is your assessment of the orthogonalization technique used by qr?

10. Let A=orth(rand(6)) and B=rand(6). Check

```
norm(A*B,2)=norm(B)
cond(A*B)=cond(B).
```

COMMENTS

See *The QR Factorization* for other orthogonalization methods.

The QR Factorization

ABSTRACT

Every matrix A can be written as QR where Q is an orthogonal matrix and R is an upper triangular matrix. This factorization can be used to find the solution to the least squares problem.

MATLAB COMMANDS

```
qr, null, orth
```

LINEAR ALGEBRA CONCEPTS

Householder Matrix, Givens Matrix, Orthogonal Matrix, Least Squares Solution, Column Space, Null Space, Orthogonal Complement

BACKGROUND

This topic can be approached by considering the Gram Schmidt Orthogonalization algorithm, but instead we are going to consider the orthogonalization methods of Householder and Givens. We begin with the Householder method.

Householder Orthogonalization

Recall the Householder matrix $H = H(w) = I_n - (\frac{2}{w^T w})ww^T$ where w is an $n \times 1$ column vector from page 33. Observe that $Hx = x - 2\text{proj}_w(x)$, where $\text{proj}_w(x)$ is the projection of x onto w, and from *Graphics*, you know that this is the reflection of x perpendicular to the vector w (see page 44.) The Householder matrices are symmetric, orthogonal, and self-inverse.

Theorem 1. *Let H be a Householder matrix, then $H = H^T = H^{-1}$.*

We can build a Householder matrix in MATLAB with any vector.

```
v=rand(5,1);
H=eye(5)-(2/(v'*v))*(v*v')
H-H'
H*H
```

Householder matrices can be used in a triangularization method by applying the following theorem. The proof tells how to select the vector w promised by the conclusion.

Theorem 2. *Given vectors x and y with $\|x\| = \|y\|$ there is a vector w such that for $H = H(w)$, $Hx = y$.*

PROOF: Let $w = x - y$ and $H = H(w)$. We want to show that $Hx = y$. Let $w_1 = \text{proj}_w(x)$ and $w_2 = \text{proj}_w(-y)$, then $w_1 \cdot (x - w_1) = 0$ and $w_2 \cdot (-y - w_2) = 0$. By the Pythagorean Theorem

$$\|x\|^2 = \|w_1\|^2 + \|x - w_1\|^2$$
$$\|y\|^2 = \| - y\|^2 = \|w_2\|^2 + \| - y - w_2\|^2$$

163

Now since $\|x\| = \|y\|$ and x and $-y$ are the adjacent sides of a rhombus with $x - y$ on the diagonal, we get $\|w_1\|^2 = \|w_2\|^2$ and $\|x - w_1\| = \|-y - w_2\|$. Thus $x - y = w_1 + w_2 = 2w_1$ and so $y = x - 2\text{proj}_w(x) = Hx$. ∎

Given a vector v suppose we want to select w so that for the Householder matrix $H = H(w)$

$$Hv = \begin{bmatrix} r \\ 0 \\ \vdots \\ 0 \end{bmatrix}$$

where r is to be determined. Notice that to satisfy the hypothesis of Theorem 2 we need $\|v\| = \|re_1\| = |r|$ and so we choose $w = v - re_1$. Let's try this in MATLAB.

```
v=rand(5,1); e1=zeros(5,1); e1(1)=1;
r=norm(v), w=v-r*e1;
H=eye(5)-(2/(w'*w))*w*w'
H*v-r*e1
```

This works just fine unless v is close to re_1 and then we admit the possibility of round-off error (see page 67 of *Errors.*) In this case we choose $r = -\|v\|$ and avoid the problem while still retaining the feature that $|r| = \|v\| = \|re_1\|$.

Theorem 3. *The QR Factorization. Let A be any matrix, then there is an orthogonal matrix Q and an upper triangular matrix R such that $A = QR$.*

PROOF: If the first column of A is all 0's, go on to the second column, otherwise choose a Householder matrix H_1 so that

$$H_1 A = \begin{bmatrix} * & * & \cdots & * \\ 0 & * & \cdots & * \\ \vdots & & & \vdots \\ 0 & * & \cdots & * \end{bmatrix}$$

Now choose

$$H_2 = \begin{bmatrix} 1 & 0 & \cdots & 0 \\ 0 & & & \\ \vdots & & H & \\ 0 & & & \end{bmatrix}$$

where H is chosen to zero out the column below the $(1, 1)$ entry of $H_1 A(2 : n, 2 : n)$. Continue this process to get

$$H_{n-1} \ldots H_2 H_1 A = R$$

and let $Q = (H_n \ldots H_1)^T$. ∎

We can do this easily for a 3×3 matrix in MATLAB.

```
A=rand(3); v1=A(:,1); e1=[1;0;0];
r1=norm(v1); w1=v1-r1*e1;
H1=eye(3)-(2/(w1'*w1))*(w*w');
A1=H1*A
A2=A1(2:3,2:3)
v2=A2(:,1); e2=[1;0];
r2=norm(v2); w2=v2-r2*e2;
H2=eye(2)-(2/(v'*v))*(v*v');
A3=H2*A2
H3=eye(3); H3(2:3,2:3)=H2
R=H3*H1*A
Q=H1*H3
Q*R-A
```

MATLAB has a function qr which does this factorization.

```
[Q,R]=qr(A)
Q*R-A
```

This factorization has a number of very nice features.

Orthonormal Bases

Suppose that A is $m \times n$ with $m \geq n$ and rank$(A) = n$. We apply the QR factorization to A to get $A = QR$ where Q is $m \times m$ and R is $m \times n$. Let $A = [\, v_1 \;\; \cdots \;\; v_n \,]$, $Q = [\, w_1 \ldots w_n \ldots w_m \,]$, and

$$R = \begin{bmatrix} R_1 \\ 0 \end{bmatrix} = [\, u_1 \;\; \cdots \;\; u_m \,],$$

then we see that $v_i = Qu_i$ for $i = 1, \ldots, n$. Since each u_i has zeros in coordinates $n+1, \ldots, m$, v_i is a linear combination of $w_1, \ldots, w_n$. Thus

$$\text{span}(v_1, \ldots, v_n) \subseteq \text{span}(w_1, \ldots, w_n)$$

and since both have dimension n,

$$\text{span}(v_1, \ldots, v_n) = \text{span}(w_1, \ldots, w_n).$$

It follows that $w_1, \ldots, w_n$ is an orthonormal basis for the Column Space of A. This is how MATLAB's orth finds a basis for the Column Space of A.

```
A=magic(8); r=rank(A);
[Q,R]=qr(A)
Q(:,1:r)
```

```
orth(A)
```

Now look at the remaining columns of Q, $w_{n+1}, \ldots, w_m$ and observe that each of these w_i's is in the orthogonal complement of the Column Space of A and again by dimension, we get that $w_{n+1}, \ldots, w_m$ is a basis for the perp of the Column Space of A. From *Projections* we know that the perp of the Column Space of A^T is the Null Space of A. Thus we can get an orthonormal basis for the Null Space of A by applying the QR factorization to A^T.

```
[Q,R]=qr(A');

N=Q(:,r+1:8)

A*N
```

MATLAB's `null` function uses a more elaborate version of the QR factorization.

Condition Number

If A is a square matrix and $A = QR$ we have $\|A\|_2 = \|R\|_2$ by Theorem 5 of *Orthonormal Bases*, and $\|A^{-1}\|_2 = \|R^{-1}Q^T\|_2 = \|R^{-1}\|_2$ and so $\text{cond}_2(A) = \text{cond}_2(R)$. Thus the QR decomposition does not change the condition number.

```
A=rand(5)+100*diag(rand(5,1))

[Q,R]=qr(A)

norm(A), norm(R)
```

Least Squares

Suppose that A is an $m \times n$ matrix where $m \geq n$ and $\text{rank}(A) = n$. We want a least squares solution to the overdetermined system $Ax = b$. Let $A = QR$ where Q is $m \times m$ and R is $m \times n$. Notice that the residual

$$\|r\| = \|Ax - b\|_2 = \|Q^T Ax - Q^T b\|_2 = \|Rx - Q^T b\|_2.$$

If we let $R = \begin{bmatrix} R_1 \\ 0 \end{bmatrix}$ where $R1$, is $n \times n$ and $Q^T b = \begin{bmatrix} u \\ v \end{bmatrix}$ where u is $n \times 1$, then

$$\|r\|^2 = \|R_1 x - u\|_2^2 + \|v\|_2^2.$$

Since v is independent of x and $\text{rank}(A) = \text{rank}(R) = \text{rank}(R_1)$, we can solve $R_1 x = u$ so that the minimal residual is $\|v\|_2$ and the least squares solution is $x = R_1^{-1} u$.

```
A=magic(8); r=rank(A); A=A(:,1:r),

b=ones(8,1);

[Q,R]=qr(A),

R1=R(1:r,:), c=Q'*b,

x=R1\c(1:3)
```

MATLAB's \ computes the least squares solution when $m \geq n$.

```
A\b
```

Givens Orthogonalization

A *Givens rotation,* (see page 'givens',) $G = \text{givrot}(n, i, j, c, s)$, where $c^2 + s^2 = 1$ is arrived at by starting with an identity matrix $G = I$ and assigning $G(i, i) = c, G(i, j) = s, G(j, i) = -s$, and $G(j, j) = c$. This produces a rotation in the ij−plane.

Theorem 4. $G = givrot(n, i, j, c, s)$ *is an orthogonal matrix.*

The method is based on the fact that a Givens rotation affects only entries in rows i and j.

Theorem 5. *For any vector v and integers $i < j$ there is a Givens rotation G such that for* $w = Gv$, $w(j) = 0$ *and* $w(k) = v(k)$ *for all* $k \neq i, j$.

PROOF: Since Gv changes only the i and j entries, we get $w(k) = v(k)$ for $k \neq i, j$. To get $w(j) = 0$ we simply need to solve these equations for c and s

$$-sv(i) + cv(j) = 0 \qquad c^2 + s^2 = 1.$$

Let

$$s = \frac{v(j)}{\sqrt{v(i)^2 + v(j)^2}} \quad \text{and} \quad c = \frac{v(i)}{\sqrt{v(i)^2 + v(j)^2}}$$

∎

Suppose that we have

```
v=rand(5,1)
```

and we want to zero out the entry in the third position using the entry in the second position.

```
d=sqrt(v(2)^2 + v(3)^2)
s=v(3)/d' c=v(2)/d,
G=eye(5); G(2,2)=c; G(2,3)=s;
G(3,2)=-s; G(3,3)=c;
G*v
v-G*v
```

You notice that only the entries in the second and third positions have been affected. In this respect, the Givens method is similar to Gaussian Elimination.

Say we want to zero out $A(j, k)$. We can do this with a Givens rotation G so that GA differs from A only in the ith and j^{th} rows for any $i < j$. Let $w = A(:, k)$ and choose G as in Theorem 5.

Now by successively choosing $n - 1$ Givens rotations zero out below the $(1, 1)$ entry. Continue on to the second column by choosing $n - 1$ Givens matrices until we zero out below $(2,2)$. Finally an upper triangular matrix is obtained. It is inadvisable to actually multiply out these Givens rotations since that will lead to an excessive number of arithmetic operations. Since Givens rotations act on two rows, they can be multiplied much more economically than by full matrix multiplication.

```
A=magic(5); B=A
G*A
B(2,:)=G(2,2)*A(2,:)+G(2,3)*A(3,:)
B(3,:)=G(3,2)*A(2,:)+G(3,3)*A(3,:)
G*A-B
```

PROBLEMS

1. Use MATLAB's qr function which produces a QR Decomposition on the following matrices

    (1) A=list(5); A=A(:,1:2); (2) B=magic(6); B=B(:,1:5);

Compare the rank(A) with the rank(R). What about norm? The matrix list is from page 34.

2. In each of the matrices in problem 1 find a basis for the Column Space and the Null Space using the QR Decomposition.

3. Let A=rand(5) and A=Q*R be given by the QR Decomposition. Compare A'*A with R'*R. Explain why this happens.

4. Find the least squares solution to $Ax = b$ where

    (1) A=magic(8), b=(1:8)'

    (2) A=list(8), b=[1 -1 1 -1 1 -1 1 -1]'

    (3) A=V(:,1:4) where V=vander(1:6), b=[0 1 0 1 0 1]

5. Write a MATLAB function [ls,res]=qrlesq(A,b) which produces the least squares solution to Ax=b using the QR factorization. ls is the least squares solution and res is the minimum residual. You may use MATLAB's qr in this function. Check your qrlesq on the problems in 4.

6. Let A=magic(4) and perform a QR factorization using Householder matrices as in the Background discussion above.

7. Write a MATLAB function [Q,R]=houseqr(A) which computes a QR factorization using Householder reflections. Test your houseqr on the matrices in problem 1 and magic(4).

8. The Givens rotations can be used to selectively to zero a single entry of a matrix. Let A=triu(ones(3),-1)+diag(1:3) and use two Givens rotations to do a QR factorization of A.

9. Write a MATLAB function [Q,R]=triqr(A) which finds a QR factorization of a tridiagonal matrix A using Givens' rotations. Test your triqr on A=triu(tril(rand(5),1),-1). What does the residual Q*R-A look like?

Eigenvalues

ABSTRACT

We consider when and how a matrix can be shown to be equivalent to a diagonal matrix. The fundamental notions of "eigenvalue" and "eigenvector" are developed.

MATLAB COMMANDS

```
eig, plot, inv, poly, roots, rank
```

LINEAR ALGEBRA CONCEPTS

Eigenvalue, Eigenvector, Eigenspace, Diagonalizable, Characteristic Polynomial, Similarity

BACKGROUND

A matrix A is ***diagonalizable*** if there is an invertible matrix B and a diagonal matrix D such that $B^{-1}AB = D$. If we let $D = \text{diag}(\lambda_1, \ldots, \lambda_n)$ and $B = [v_1, \ldots, v_n]$, then

$$AB = [Av_1, \ldots, Av_n] = [\lambda_1 v_1, \ldots, \lambda_n v_n] = BD.$$

Thus $Av_i = \lambda_i v_i$ for $i = 1, \ldots, n$. We say λ is an ***eigenvalue*** and $v \neq \vec{0}$ is an ***eigenvector*** if $Av = \lambda v$. From these observations the diagonal entries of D are seen to be eigenvalues and the columns of B are eigenvectors.

The eigenvectors and the eigenvalues are important notions independently of diagonalization. The eigenvectors give "direction" to a matrix. The following program `eigplot.m` should be typed in for use in the rest of this discussion. The function `eigplot(A)` plots 2-dimensional vectors x which lie on regularly spaced intervals on the unit circle. After each x is plotted, A*x is computed and plotted. It will pause after each pair x and A*x are plotted waiting for you to type any key to prompt it to continue.

```
function eigplot(A)

% eigplot(A). Plots x and A*x.
% in 16 evenly spaced intervals
% on the unit circle.

clg, t=0:2*pi/16:2*pi;
axis([-2,2,-2,2]); hold
x=cos(t); y=sin(t); plot(x,y);
for i=t;
   x=cos(i); y=sin(i); z=A*[x;y];
   plot([0;x],[0;y]);
   plot([0;z(1)],[0;z(2)]);
   pause
end
hold
```

Try this in MATLAB

```
A=[4,2;2,4]
eigplot(A)
```

The first thing you should notice is the tendency for the Ax vectors to cluster. While the x's are evenly space around the interval, the Ax's are not. This is what I mean by the "direction" of a matrix. Now run this program again but watch closely for those times when Ax is plotted over x. That will happen when Ax is a scalar multiple of x, in other words $Ax = \lambda x$ for some λ. You should see this happen four times. If not, run the progam again. There are only two eigenvalues for a 2×2 matrix. The eigenvectors and eigenvalues for

$$A = \begin{bmatrix} 4 & 2 \\ 2 & 4 \end{bmatrix}$$

are

$$\lambda_1 = 6 \quad v_1 = \begin{bmatrix} 1 \\ 1 \end{bmatrix} \qquad \lambda_2 = 2 \quad v_2 = \begin{bmatrix} 1 \\ -1 \end{bmatrix}$$

But look! If $Av_1 = 6v_1$, then $A(-v_1) = -Av_1 = -6v_1 = 6(-v_1)$, so that $-v_1$ is also an eigenvector. Every vector on the line passing through v_1 is an eigenvector. Suppose that we have $Ax = \lambda x$, then $\lambda x - Ax = \vec{0} = (\lambda I_n - A)x$ and so x is in the Null Space of $\lambda I_n - A$. This null space is called the **eigenspace of** λ. Going back to our program, `eigplot`, try

```
B=rand(2); A=2*B'*B; eigplot(A)
```

How many eigenvectors can you find? Try

```
A=[0,-1;1,0]; eigplot(A)
```

Do not be surprised, this is a rotation matrix. There are eigenvalues and eigenvectors for this A but they are complex numbers and complex vectors. You can see them using the MATLAB function

```
[P,D]=eig(A)
```

The diagonal entries of D are the eigenvalues and the columns of P are the eigenvectors. Now try

```
A=[2,2;0,2]; eigplot(A)
```

This time you will notice that there is only one line of eigenvectors, rather than two. This last example is important since A cannot be diagonalized. What happens with

```
[P,D]=eig(A)
```

Notice that $\lambda = 2$ is repeated twice as an eigenvector and that P has two linearly dependent columns and hence P is not invertible. We are ready for the next theorem.

Theorem 1. *A is diagonalizable if and only if there is a basis for* R^n *of eigenvectors.*

Our example A=[2,2;0,2] is not diagonalizable. If we try to solve Ax=λx for λ we get only one eigenvalue $\lambda = 2$, but the Null Space of $2I - A$ has dimension 1, and so there can be no basis for R^2 of eigenvectors. Determining quickly if a matrix is diagonalizable is not so easy. The next theorem is a test that works frequently.

Theorem 2. *If the eigenvalues* $\lambda_1, \ldots, \lambda_n$ *for A are distinct, then A is diagonalizable.*

Suppose that we try

```
A=rand(5); eig(A)
```

The eigenvalues are distinct and so A is diagonalizable.

When we have $(\lambda I - A)v = \vec{0}$ and $v \neq \vec{0}$, $\lambda I - A$ is singular, yielding $\det(\lambda I - A) = 0$. If we replace λ by x, $\det(xI - A)$ is a polynomial called the ***characteristic polynomial*** of A. The eigenvalues of A are the roots of the characteristic polynomial. The MATLAB function f=poly(A) will find the characteristic polynomial of A. Follow this with roots(f), and you will have the eigenvalues of A. See the Comments at the end of this project about this approach to eigenvalues. We say that A is ***similar*** to C if there is an invertible B such that $B^{-1}AB = C$. In this terminology, A is diagonalizable if A is similar to a diagonal matrix.

Theorem 3. *If A is similar to C, then A and C have the same characteristic polynomial, and thus they have the same eigenvalues.*

In MATLAB try

```
A=magic(5); P=rand(5)
C=inv(P)*A*P
eig(A)
eig(C)
```

Theorem 4. *If A is a triangular matrix, then the eigenvalues of A are the diagonal entries of A.*

The ***multiplicity of*** λ in the characteristic polynomial is m if $(x - \lambda)^m$ divides the characteristic polynomial but $(x - \lambda)^{m+1}$ does not. In other words, $x - \lambda$ occurs as a factor exactly m times. There is an important connection between the multiplicity of λ and the eigenspace of λ. First try this in MATLAB

```
A=[[2*eye(2);zeros(2)],ones(4,2)]
[P,D]=eig(A)
```

We see that the multiplicity of the eigenvalue $\lambda = 2$ is 3. We also note that the eigenvectors appearing in P are not linearly independent. We can check this easily with

```
rank(P)
```

Finally we can compute the dimension of the eigenspace for $\lambda = 2$

```
4-rank(2*eye(4)-A)
```

The next theorem includes a proof since the proof is simple and rarely appears in textbooks.

Theorem 5. *If λ has multiplicity m, then the dimension of the eigenspace of λ is $\leq m$.*

PROOF: Let $v_1, \ldots, v_k$ be a basis for the Null Space of $\lambda I - A$. We want to show that $m \geq k$. Expand $v_1, \ldots, v_k$ to a basis for $\mathbb{R}^n$, $v_1, \ldots, v_k, v_{k+1}, \ldots, v_n$, and let $B = [v_1, \ldots, v_n]$. Let $C = B^{-1}AB$.

We will now get a closer look at the first k columns of C.

$$B^{-1}[Av_1, \ldots, Av_k] = [B^{-1}\lambda v_1, \ldots, B^{-1}\lambda v_k]$$
$$= [\lambda B^{-1}v_1, \ldots, \lambda B^{-1}v_k]$$
$$= [\lambda e_1, \ldots, \lambda e_k]$$

Thus the first k columns look like

$$\begin{bmatrix} \lambda & 0 & \cdots & 0 \\ 0 & \lambda & \ddots & \vdots \\ \vdots & & \ddots & 0 \\ 0 & \cdots & 0 & \lambda \\ 0 & \cdots & \cdots & 0 \\ \vdots & & & \vdots \\ 0 & \cdots & \cdots & 0 \end{bmatrix}.$$

But now we see that $(x - \lambda)^k$ divides the characteristic polynomial of C which, from Theorem 3, is the same as the characteristic polynomial of A. Thus $m \geq k$. ∎

This offers a way of definitively checking if a matrix is diagonalizable.

Theorem 6. *A is diagonalizable if and only if for each eigenvalue λ the multiplicity of λ is the dimension of the eigenspace of λ.*

The matrix A constructed above is not diagonalizable since the multiplicity of the eigenvalue $\lambda = 2$ is 3, while the dimension of the eigenspace is 2. When a matrix is diagonalizable, we would like to treat it just like a diagonal matrix. Unfortunately, some translation is necessary to get the desired results. The next theorem shows how to do this when computing powers.

Theorem 7. *Suppose that A is diagonalizable with $P^{-1}AP = D$, $D = \text{diag}(\lambda_1, \ldots, \lambda_n)$ and $P = [v_1, \ldots, v_n]$. If x is any vector and c is the vector of coordinates of x with respect to $v_1, \ldots, v_n$, that is, $x = Pc$, then for all m*

$$A^m x = PD^m c = \sum_{i=1}^{n} c_i \lambda_i^m v_i$$

PROBLEMS

1. Try eigplot on A=[4,6;6,4]. Can you find the eigenvectors? Use MATLAB's [P,D]=eig(A) to find the eigenvectors. Put a hold on your eigplot graph and use vecplot (see page 43 of *Graphics*) to plot the eigenvectors on top of the eigplot.

2. Determine which of the following matrices are diagonalizable. See *Building Matrices* for the definitions of these matrices.

    (1) `pascal(5)`
    (2) `compan([1 -5 10 -10 5 -1])`
    (3) `list(5)`
    (4) `jord(3,5)`
    (5) `ele1(5,2,5)`
    (6) `ele3(5,2,3,4)`
    (7) `ones(5)`
    (8) `B'*B` where `B=rand(5)`
    (9) `house((1:5)')`

3. For each of the matrices in problem 2 compute `prod(eig(A))`, `det(A)`, and `poly(A)`. What can you say about these answers? Can you explain why this is so?

4. When a linear system represented by a matrix A is used to model a process, Ax represents the output of the process on input x. We would like to know the outcome of the process after it has been repeated m times. To do this we compute A^m. Let `B=rand(5)`, `A=B'*B` and `x=ones(5,1)`. Compute `A^100*x` and then use Theorem 7. Which method uses more flops? For discussion of flops see *Flops*. Be sure to count the flops involved in computing `eig`.

5. When A is invertible $Ax = b$ can be solved with $x = A^{-1}b$. If A is an invertible diagonal matrix, we can go further to say $x(i) = b(i)/A(i, i)$. Suppose that A is diagonalizable, so that $P^{-1}AP = D$ where D is diagonal and $Ax = PDP^{-1}x = b$. Let $z = P^{-1}x$ be new variables. Then to solve $PDz = b$, multiply both sides by P^{-1}, so $Dz = P^{-1}b$, a diagonal system, which is easily solved. Now x is obtained by $x = Pz$. Let `A=hilb(5)` and find an invertible matrix P and a diagonal matrix D such that

    `inv(P)*A*P=D.`

Solve the matrix equation `Ax=b` where `b=(1:5)'` using the technique described above.

6. Compute `eig(house(rand(n,1)))` for n=2,3,4,5,6. Can you explain what is going on here?

7. Let `A=rand(5,3)` and `P=A*inv(A'*A)*A'`. Compute

    `rank(A)`
    `rank(P)`

Can you explain why this happens? Now what is `eig(P)`? Try some other matrices P based on different selections of A, for example `A=rand(4,2)`. Can you explain what is happening?

8. MATLAB computes the characteristic polynomial of a matrix A with the function `poly(A)` from the eigenvalues. You can mimic MATLAB's computation with the function `expan` (see page 38.) Let `A=rand(10)`, `r=eig(A)'`, `p=expan(r)`, and `q=poly(A)`. Compare these polynomials with `max(abs(p-q))`.

9. Consider the matrix `A=[5/2 , 1/2 ; 1/2 , 5/2]`. Plot the circle $x^2 + y^2 = 1$ and transform it as follows:

    `t=0:pi/20 :2*pi; x=cos(t); y=sin(t);`

```
z=A*[x;y]; x1=z(1,:); y1=z(2,:)
```

Now plot the transformed circle together with the original circle.

```
plot(x1,y1), hold, plot(x,y)
```

Find the eigenvectors of A and plot them using vecplot. Can you make a connection between the eigenvectors and the plots?

10. Write a MATLAB function b=diagnble(A) which returns b=1 if A is diagonalizable and b=0 otherwise.

11. Test your diagnble on the matrices in problem 2.

12. Let A=rand(5,3) and P=A*inv(A'*A)*A' the projection matrix. Compute

```
eig(P)
```

```
sum(eig(P))
```

```
rank(P)
```

```
rank(A)
```

What happens? Start with another A, does it happen again? Try to explain what is happening.

13. Let A=rand(5) and B=rref(A). Is there any relationship between the eigenvalues of A and the eigenvalues of B?

14. Here is a simple way to compute the largest eigenvalue of a matrix. For more information on this method see *The Power Method*. Let A=magic(5) and just to have a peek at the eigenvalues

```
eig(A)
```

Now choose a random initial vector x=rand(5,1) and apply A to x iteratively with the following program.

```
for i=1:20, y=x; x=A*y; x./y, end
```

Notice how the vector x./y is converging to the largest eigenvalue of A. Try this with some other matrices.

15. This problem uses polynomials. For general infromation on how MATLAB handles polynomials see *Polynomials*. If $f(x) = a_1x^{n-1}+a_2x_{n-2}+\ldots+a_{n-1}x+a_n$ is a polynomial and A is a square matrix, then it is possible to substitute A for x to get $f(A) = a_1A^{n-1} + a_2A^{n-2} + \ldots + a_{n-1}A + a_nI$. The MATLAB function polyvalm does this. The Cayley-Hamilton Theorem says that if $f(x)$ is the characteristic polynomial of A, then $f(A) = 0$. While this theorem is true for all square matrices, it is easy to see for diagonalizable matrices, since for $P^{-1}AP = D$, $f(A) = f(PDP^{-1}) = Pf(D)P^{-1}$. But look at $D = \mathrm{diag}(\lambda_1,\ldots,\lambda_n)$, $f(D) = \mathrm{diag}(f(\lambda_1),\ldots,f(\lambda_n)) = \mathrm{diag}(0,\ldots,0) = 0$. Try this using MATLAB's poly and polyvalm on A=rand(5). Now let g be a row vector which represents the polynomial x^{20}, (g should be a 1×21 matrix,) and suppose that f=poly(A). Let [q,r]=deconv(g,f). Compare A^20, polyvalm(g,A), and polyvalm(r,A).

16. This exercise requires some Calculus. A system of linear differential equations has the form

$$a_{11}y_1 + a_{12}y_2 + \cdots + a_{1n}y_n = y_1'$$
$$a_{21}y_1 + a_{22}y_2 + \cdots + a_{2n}y_n = y_2'$$
$$\vdots$$
$$a_{n1}y_1 + a_{n2}y_2 + \cdots + a_{nn}y_n = y_n'$$

where we wish to find solutions $y_1 = f_1(t), \ldots, y_n = f_n(t)$. We can turn this into a matrix equation by letting A be the matrix of coefficients a_{ij} and letting $\mathbf{y} = [y_1, \ldots, y_n]^T$ and defining $\mathbf{y}' = [y_1', \ldots, y_n']^T$, so that the system becomes $A\mathbf{y} = \mathbf{y}'$. We know how to solve an equation of the form $y' = ay$ from Calculus with $y = ce^{at}$, where c is an integration constant which needs to be determined from input data. Thus, if A is a diagonal matrix, we can easily solve the system $A\mathbf{y} = \mathbf{y}'$. Suppose now that A is diagonalizable, with $P^{-1}AP = D$. Define new variables by $\mathbf{z} = P^{-1}\mathbf{y}$ so that $\mathbf{z}' = P^{-1}\mathbf{y}'$. Notice that

$$A\mathbf{y} = \mathbf{y}' \iff PDP^{-1}\mathbf{y} = \mathbf{y}' \iff PD\mathbf{z} = P\mathbf{z}' \iff D\mathbf{z} = \mathbf{z}'.$$

Thus we need only solve for $\mathbf{z}$ and translate the answer back by $\mathbf{y} = P\mathbf{z}$. Now if we have known initial conditions, say $\mathbf{y}(0) = b$, then we can also solve for the unknown integration constants, since $b = \mathbf{y}(0) = P\mathbf{z}(0)$. Use this technique to solve the following system of differential equations:

$$2y_1 + 5y_2 + 2y_3 = y_1'$$
$$y_1 + 8y_2 + 4y_3 = y_2'$$
$$2y_1 + 4y_2 + y_3 = y_3'$$

and initial conditions $\mathbf{y}(0) = (1 : 3)'$.

COMMENTS

The method for finding eigenvalues which is usually taught in an elementary linear algebra class is to find the characteristic polynomial and then find its roots. The MATLAB functions `poly` and `roots` allow you to use this method. This is not a realistic view of the computation since MATLAB finds the eigenvalues using an iterative method with `eig` and constructs the characteristic polynomial from the eigenvalues. To get the roots of a polynomial, $f(x)$, MATLAB forms a companion matrix of the polynomial. The eigenvalues of the companion matrix are the roots of the polynomial. Now MATLAB uses `eig` again in the function `roots`.

The Arnold Cat

ABSTRACT

In this project we have some fun with a transformation used by V.I. Arnold to study flows on the torus. This transformation provides an excellent opportunity to study eigenvalues and eigenvector behavior.

MATLAB COMMANDS

```
plot, eig
```

LINEAR ALGEBRA CONCEPTS

Eigenvalues, Eigenvectors, Linear Transformations

BACKGROUND

Arnold's Transformation

Consider the linear transformation $T : \mathrm{R}^2 \to \mathrm{R}^2$ defined by $T(v) = Av$ where

$$A = \begin{bmatrix} 2 & 1 \\ 1 & 1 \end{bmatrix}$$

V.I. Arnold used a picture of a cat to demonstrate some of the features of this transformation. We will do the same with a poor substitute for the famous *Arnold cat*. In MATLAB do the following:

```
t=0:pi/100:2*pi;
head=.5+.1*[cos(t);sin(t)];
plot(head(1,:),head(2,:))
hold
eyes=[.45,.55;.55,  .55];
plot(eyes(1,:),eyes(2,:),'+')
nose=[.5;.47];
plot(nose(1,:),nose(2,:),'o')
```

That gives you a simple "cat" to look at. We will now bundle this up into a single matrix.

```
cat=[head,eyes,nose];
hold
plot(cat(1,1:201),cat(2,1:201),cat(1,202:204),cat(2,202:204),'*')
```

That should have given you a cat. Now we will build the Arnold transformation and let it act on the cat.

```
A=[2,1;1,1]
cat=A*cat;
plot(cat(1,1:201),cat(2,1:201),cat(1,202:204),cat(2,202:204),'*')
```

You will notice that the cat has been deformed a bit. Repeat this process.

176

```
cat=A*cat;
plot(cat(1,1:201),cat(2,1:201),cat(1,202:204),cat(2,202:204),'*')
```
Poor kitty. Do it again.
```
cat=A*cat;
plot(cat(1,1:201),cat(2,1:201),cat(1,202:204),cat(2,202:204),'*')
```
We must now wonder if Arnold is a cat lover. The behavior that you have just witnessed is entirely associated with the eigenvalues of the Arnold transformation. Let's take a look at them.
```
[p,d]=eig(A)
```
Notice that there are two distinct eigenvalues and they have the relationships
$$d(1,1) > 1 > d(2,2) > 0$$
We can take a look at the eigenvectors. Use the function vecplot from page 43 of *Graphics*.
```
v1=p(:,1), v2=p(:,2), vecplot(v1)
```
If v1 has negative coordinates, turn it around with
```
v1=-v1
```
Now go back to the original kitty.
```
cat=[head,eyes,nose];
plot(cat(1,1:201),cat(2,1:201),cat(1,202:204),cat(2,202:204),'*')
hold
vecplot(v1);
```
Apply the Arnold transformation several times.
```
cat=A*cat;
clf
plot(cat(1,1:201),cat(2,1:201),cat(1,202:204),cat(2,202:204),'*')
hold
vecplot(v1)
clf
cat=A*cat;
plot(cat(1,1:201),cat(2,1:201),cat(1,202:204),cat(2,202:204),'*')
hold
vecplot(v1)
vecplot(6*v1)
clf
cat=A*cat;
plot(cat(1,1:201),cat(2,1:201),cat(1,202:204),cat(2,202:204),'*')
hold
vecplot(v1)
```

```
vecplot(16*v1)
```

The eigenvector spears the cat! This pattern is not particular to the Arnold transformation, it will happen whenever there are eigenvalues with the relationships

$$\lambda_1 > 1 > \lambda_2 > 0$$

See problem 1 for an illustration of how this works.

The Expander and the Contractor

We return now to our consideration of the Arnold transformation

$$A = \begin{bmatrix} 2 & 1 \\ 1 & 1 \end{bmatrix}.$$

From Theorem 2 of *Eigenvalues* we know that A is diagonalizable, that is, the eigenvectors of the distinct eigenvalues form a basis for $\mathbb{R}^2$. We can thus write any vector in $\mathbb{R}^2$ as a linear combination of the eigenvectors,

$$u = c_1 v_1 + c_2 v_2.$$

When we apply A $n-$ times

$$A^n u = c_1 \lambda_1^n v_1 + c_2 \lambda_2^n v_2$$

where $\lambda_1 > 1 > \lambda_2 > 0$. This illustrates Theorem 7 of *Eigenvalues*. Notice that $\lim_{n\to\infty} \lambda_1^n = +\infty$ while $\lim_{n\to\infty} \lambda_2^n = 0$. So that as n goes to infinity we see that $c_2 \lambda_2^n v_2$ goes to $\vec{0}$ and thus

$$A^n u \approx c_1 \lambda_1^n v_1$$

This explains the behavior of Arnold's cat under repeated applications of the Arnold transformation. The eigenvector associated with λ_1 is expanding, and the eigenvector associated with λ_2 is contracting. The expanding eigenvector tends to pull all vectors which are not scalar multiples of v_2 in the direction of v_1. This is the heart of the idea behind the *power method* for computing eigenvalues (see *The Power Method*.)

The eigenvector, v_1, is an attractor, since every vector which is not a scalar multiple of v_2 is drawn to it. The next program will give you some idea of how powerful this attractor is. Two points are chosen at random. The Arnold transformation is applied to each. The program `catdist.m` calculates the distance between them and compares the plot of the distance with the plot of the exponential function.

```
function catdist(n)

% catdist(n). Two points are chosen at random.
% The Arnold transformation is applied to both
% of them n times. After each application the
% distance is computed. The plot is compared to
% a plot of the exponential function 2^n.

a=[2,1;1,1];
v=rand(2,1); u=rand(2,1);
dist=norm(u-v);
for i=1:n
  v=a*v; u=a*u;
  dist=[dist,norm(u-v)];
end
clf
plot(2.^(1:n+1)), hold
pause(5), plot(dist)
```

Try this a few times to get a sense for what is going on. Since the initial points are chosen at random there will be variation in the plot of the distances. Of course, the exponential function 2^n will plot the same way every time. Notice that the program will pause for a few seconds between plots.

```
catdist(10)
```

The Transformation on the Torus

You may recall the wrapping function from Trigonometry. We have been using it in this project to construct the head of the cat. The wrapping function is $(\cos(t), \sin(t))$ where t is a real number. The effect of the wrap is to overlay the values in periods of 2π. Another way to do this is by rounding down to the nearest integer using the function mod from page 16 of the *MATLAB Tutorial.*

```
t=0:.01:3;
x=mod(t,1);
plot(t,x,'.')
ax=[0,3,-2,2]; axis(ax)
```

Now you can compare this with $\sin(t)$

```
hold
plot(t,sin(t))
```

The function mod is not a continuous function, but because of the periodicity, it will get the job done. Try this

```
clf
```

```
plot(x,x)
```

If you want it to look more like a circle, it can be opened up a bit by squaring.

```
plot(x,x.^2)
```

The *torus* is a three-dimensional object which looks like a doughnut or a bagel. We can tear the torus appart with a pair of sissors. First just cut right through the torus to get a cylinder.

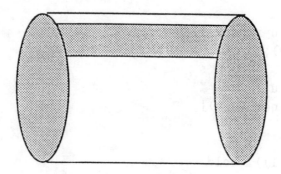

You should think of the torus as just a skin; it is not a solid doughnut. So the shaded part is empty space. Now take your sissors and cut along the top of the cylinder. As you unfold the torus, you will see a rectangle. Well, close enough. Mathematicians think of a torus as a rectangle where the opposite sides have been taped together. Fortunately, this is very easy to do algebraically with the mod function. We will form the line $y = 2x$ and wrap it around the torus using the mod function.

```
t=-10:.01:10;
x=t; y=2*t;
clf
plot(x,y,'.')
hold
x1=mod(x,1);y1=mod(y,1);
plot(x1,y1,'.')
```

You can see the line being wrapped around the torus in this plot, if you keep in mind that the top boundary is to be taped to the bottom boundary and the left boundary is to be taped to the right boundary.

Since the Arnold transformation, A, has integer entries, multiplying an integer vector by it will produce an integer vector. Keep in mind that a point u will be represented on the torus by mod(u, 1). This means that $u = v + $ mod(u, 1) where v is an integer vector. Now $Au = Av + A$ mod(u, 1), and since Av is an integer vector, it follows that Au and A mod(u, 1) will be represented on the torus by the same point, namely mod(Au, 1). This is a very important (and subtle) point in working on the torus. To see what can go wrong look at problem 5.

The following program `arnold.m` will map any data matrix d to the torus and apply the Arnold transformation to it. It uses the cat's head as the default value for d. For simplicity we will forget about the eyes and nose of the cat.

```
function arnold(n,d,A)

% arnold(n,d,A). This will plot n iterations
% of the tranformation A applied to the data
% matrix d. The default values are n=5,
% d=Arnold cat, A=Arnold transformation.

if nargin < 3, A=[2,1;1,1]; end
if nargin < 2,
  t=0:pi/500:2*pi;
  d=.5+.1*[cos(t);sin(t)];
end
if nargin < 1, n=5; end

for i=1:n
  clg
  plot(d(1,:),d(2,:),'.')
  axis([0,1,0,1])
  pause(5)
  d=A*d;
  d=mod(d,1);
end
```

PROBLEMS

1. Form a matrix with eigenvalues in the same relationship as the Arnold transformation.

```
p=rand(2);
B=inv(p)*diag([4,.5])*p;
[p,d]=eig(B)
v1=p(:,1);
```

Now form the cat as in the Background discussion.

```
cat=[head,eyes,nose];
plot(cat(1,1:201),cat(2,1:201),cat(1,202:204),cat(2,202:204),'*')
hold
```

Apply this transformation several times to the cat, plotting it with the eigenvector to see that the eigenvector will spear the deformed cat.

2. Look at the line $y = .5$ on the torus. This is built by

```
x=0:.01:1;
y=.5*ones(size(x))
```

Now apply the Arnold tranformation to this several times to see what it looks like on the torus. Try the same thing with the line $x = .5$.

3. Look at a single point on the torus and see what the Arnold transformation does to it.

```
pt=[pi/4;pi/4];
pt=A*pt;
plot(pt(1),pt(2),
hold
```

Repeat this many times.

```
for i=1:30,
pt=A*pt; pt=mod(pt,1);
plot(pt(1,:),pt(2,:))
end
```

Now try this with these points. $(.5, .5)$, $(.1, .2)$, $(1/3, 1/3)$, $(.1, .1)$, $(1/7, 6/7)$

4. Find the contracting vector for the Arnold transformation, by

```
[P,D]=eig(A); v2=P(:,2);
```

Now apply the Arnold transformation to it repeatedly.

```
x=v2;
for i=1:15, x=A*x, end
```

Now if u is any vector you can write it as

$$u = c_1 v_1 + c_2 v_2$$

for some scalars c_1, c_2 where v_1, v_2 are the eigenvectors. We know that

$$A^n u = c_1 \lambda_1^n v_1 + c_2 \lambda_2^n v_2$$

which can be written as the matrix equation

$$P \begin{bmatrix} c_1 \lambda_1^n \\ c_2 \lambda_2^n \end{bmatrix} = A^n u$$

where $P = [v_1, v_2]$. Using MATLAB we can watch what happens to the eigenvalues by solving the equation

$$Px = A^n u$$

```
u=rand(2,1);
x=P\u, d=x;
for i=1:15, u=a*u; x=P\u; d=[d,x]; end
d
```

5. Look at the points (.5,.5), (1.5,1.5),(2.5,2.5) on the torus. Show that Ax is the same point when x is allowed to take the value of each of these points. Consider now the matrix formed by

```
P=rand(2); B=P*A*inv(P);
```

The matrix B has the same eigenvalues as A, but what happens to the points Bx as x varies through these points.

6. Define

$$A_k = \begin{bmatrix} k^2 + 1 & k \\ k & 1 \end{bmatrix}$$

Notice that A_1 is the Arnold transformation. What does the Arnold cat look like on the torus after applying A_2 and A_3 instead of A_1?

7. Define

$$B_k = \begin{bmatrix} k^2 & k \\ k & 1 \end{bmatrix}$$

Repeat exercise 6 for the matrices B_2 and B_3. Do an eigenvalue analysis on the B_k's. Is there an important difference between the B_k's and the A_k's?

The Power Method

ABSTRACT

We consider some simple iterative methods for computing eigenvalues.

MATLAB COMMANDS

```
inv, *, ', norm
```

LINEAR ALGEBRA CONCEPTS

Eigenvalue, Eigenvector, Gershgorin Disc

BACKGROUND

We begin by looking at the easiest method of finding an eigenvalue, the **power method.** Given a square matrix A and a vector v consider the sequence given by $v_0 = v$ and $v_{n+1} = Av_n$. This sequence of vectors will tend towards an eigenvector, but we need to worry about the terms in the sequence growing toward overflow or underflow. Thus, at each stage we normalize, that is we assign

$$v_n = \frac{v_n}{\|v_n\|_\infty}$$

to get $\|v_n\|_\infty = 1$. Recall from *Norms and Condition Numbers* page 83 that $\|v\|_\infty = \max|v(i)|$. Suppose that after normalization we arrive at a stage where $v_n \approx v_{n+1}$ where $\approx$ means $\|v_n - v_{n+1}\|_2 < tol$, and tol is some small value we have chosen. Then we have found an eigenvector. This is easy to do in MATLAB.

```
A=ones(5)+5*eye(5), eig(A)
```

We can see the largest eigenvalue. Now use the power method to find it.

```
v=rand(5,1);
for i=1:20, y=v; v=A*y; v=v/norm(v,inf);
norm(v-y,2), pause, end
```

The norm has been inserted so that we can watch the vectors converge. The pause lets you see the value. To continue the iteration, hit any key. When this is done we will look for the eigenvalue.

```
A*v./v
```

Theorem 1. *If A is diagonalizable with a unique largest (in absolute value) eigenvalue λ, then v_n converges to an eigenvector v for λ.*

PROOF: Let $x = c_1\lambda u + \sum_{i=2}^n c_i\lambda_i u_i$ where $|\lambda| > |\lambda_i|$ for $i = 2, \ldots, n$. Then by Theorem 7 of *Eigenvalues*

$A^m x = c_1\lambda^m u + \sum_{i=2}^n c_i\lambda_i^m u_i$. Now divide both sides by λ^m to get

$$(1/\lambda^m)A^m x = c_1 v + \sum_{i=2}^n c_i(\lambda_i/\lambda)^m v_i.$$

Since $\lambda > \lambda_i$, $\lambda/\lambda_i < 1$ and so the terms in the sum tend to 0, and thus for large m we get

$$A^m x \approx c_1 \lambda^m u.$$

So for large m, $A^m x$ and $A^{m+1} x$ are approximately scalar multiples. After normalization $A^m x \approx v_m$ and $A^{m+1} v \approx v_{m+1}$ so that v_m and v_{m+1} are positive scalar multiples both of norm value 1, and hence $v_m \approx v_{m+1}$. ∎

The hypothesis of the theorem that there be a unique largest eigenvalue is needed. Try

```
A=ones(5)-5*eye(5), v=rand(5,1); eig(A)
for i=1:20, y=v; v=A*y; v=v/norm(v,inf);
norm(v-y,2), pause, end
```

In this case you note that norm(v-y,2) does not go to 0.

The proof indicates that the starting vector x is not so critical as long as the coefficient $c_1 \neq 0$. Typically we would not know this coefficient, but for an arbitrary x it is unlikely that $c_1 = 0$.

With this approximate eigenvector, v, in hand, we need to find λ. A reasonable approach would be to look at $Av = \lambda v$. Since it may not be the case that Av is an exact multiple of v, this approach is problematic. In fact, viewing $Av = \lambda v$ as an $m \times 1$ system with variable λ, this system is likely to be inconsistent. We will use a "least squares solution" to get the eigenvalue. See *Least Squares* for a discussion of least squares solutions.

Theorem 2. *The least squares solution x to $Av = vx$ is given by $x = \frac{v^T Av}{v^T v}$. The quotient $\frac{v^t Av}{v^t v}$ is called the **Rayleigh Quotient**.*

PROOF: Look at the equation $vx = Av$. The normal equation (see page 124 of *Projections*) is $v^T vx = v^T Av$, which has the solution $x = \frac{v^T Av}{v^T v}$. ∎

We can use this in MATLAB.

```
A=ones(5)+5*ones(5);
x=rand(5,1)
for i=1:20, y=v; v=A*y; v=v/norm(v,inf); end
```

We have an approximation to an eigenvector with v. The Rayleigh Quotient will approximate the eigenvalue.

```
r=(v'*A*v)/(v'*v)
```

Suppose that A is invertible and that $\lambda \neq 0$ is an eigenvalue of A with eigenvector v. We have $Av = \lambda v$, and multiplying by inverses $A^{-1} v = (1/\lambda)v$. If $1/\lambda$ is the largest (in absolute value) eigenvalue of A^{-1}, then λ is the smallest (in absolute value) eigenvalue of A. Thus we can find the smallest eigenvalue of A by applying the power method the the inverse of A. Notice that v is an eigenvector for both A and A^{-1}. To get an eigenvalue for A from v compute the Rayleigh Quotient on A. Try

```
A=ones(5)+diag(1:5); B=inv(A);
for i=1:20, y=v; v=B*y; v=v/norm(v,inf); end
norm(y-v,2)
r=(v'*A*v)/(v'*v)
min(eig(A))
```

Another variation is the ***shift method.*** Here you choose a scalar s and apply the power method to the shifted matrix $A - sI$. Notice that if v is an eigenvector of A with eigenvalue λ, then $(A - sI)v = Av - sv = (\lambda - s)v$, so that $\lambda - s$ is an eigenvalue of the shifted matrix and v is an eigenvector of both A and $A - sI$.

By combining the inverse method with the shift method we get an effective iteration, the ***inverse shift method.*** If you choose s close enough to λ, then $\lambda - s$ is the smallest eigenvalue of the shifted matrix $A - sI$, thus applying the power method to the *inverse* of $A - sI$ will converge to an eigenvector. The problem is choosing the shift s. Let's try this in MATLAB, but we will cheat a bit.

```
A=ones(3)+diag(1:3), eig(A)
```

Now choose s=2 and invert B=A-2*eye(3)

```
B=inv(A-2*eye(3)), x=rand(3,1);
for i=1:20, y=x; x=B*x; x=x/norm(x,inf); end
r=(x'*A*x)/(x'*x)
```

One way to choose a shift is the Rayleigh Quotient $s = \frac{x^T A x}{x^T x}$ of some vector x. This presupposes that you have vector x, in mind for the eigenvector. A less obvious choice is based on Gershgorin's Theorem. Let r_i be the sum of the entries in the ith row of A minus the diagonal element, a_{ii}. Plot each diagonal entry, a_{ii}, in the complex plane and make it the center of a circle of radius r_i. Gershgorin's Theorem says that the eigenvalues are inside the union of these circles. This suggests using the diagonal entries a_{ii} as shift values.

Theorem 3. *Gershgorin's Theorem. Let A be an $n \times n$ matrix with entries denoted by a_{ij} and define $r_i = \sum_{j \neq i} |a_{ij}|$. If λ is an eigenvalue, then there is an i such that $|\lambda - a_{ii}| \leq r_i$*

PROOF: Suppose that $Av = \lambda v$ and that $v_i = \max_j v_j$. We may assume that $v_i > 0$. Then the ith entry in Av is

$$A(i,:) \cdot v = \sum_{j=1}^{n} a_{ij} v_j = \lambda v_i.$$

Now rearranging we get

$$\lambda v_i - a_{ii} v_i = \sum_{j \neq i} a_{ij} v_i \leq \sum_{j \neq i} |a_{ij}| v_i = r_i v_i$$

Thus $|\lambda - a_{ii}| \leq r_i$. ∎

We are going to plot the Gershgorin Disc in MATLAB. Let

```
A=ones(3)+diag(1:3);
r1=sum(abs(A(1,:))-abs(A(1,1)));
r2=sum(abs(A(2,:))-abs(A(2,2)));
r3=sum(abs(A(3,:))-abs(A(3,3)));
```

Now we make the circles. Since we can see that the plots will lie inside $-1 \leq x \leq 6$ and $-2 \leq y \leq 2$. Set the axis

```
axis([-1 6 -2 2])
t=0:pi/20:2*pi; i=sqrt(-1); z=cos(t)+i*sin(t);
z1=A(1,1)*ones(z)+r1*z;
plot(z1), hold
z2=A(2,2)*ones(z)+r2*z
plot(z2)
z3=A(3,3)*ones(z)+r3*z
plot(z3)
```

Now fill in the eigenvalues.

```
plot(eig(a)+i*eps,'o')
```

We have used a little trick with eps to induce MATLAB into plotting the eigenvalues in the complex plane.

We call the circle and its interior a **Gershgorin Disc.** The theorem states that each eigenvalue falls inside a Gershgorin Disc. Unfortunately, we cannot conclude that each disc contains an eigenvalue. There is a second Gershgorin Theorem, which we will not prove, that allows us to conclude if one disc does not meet any of the other discs, then there is an eigenvalue in that disc.

PROBLEMS

1. Use the power method to find an eigenvalue for each of the following; you can check for accuracy by comparing with MATLAB's eig.

(1) `compan([6,-11,6])`
(2) `list(5)`
(3) `pascal(5)`
(4) `hilb(5)`
(5) `jord(3,3)`

What happens with `jord(3,3)`? For the definitions of some of these matrices see *Building Matrices* .

2. Write a MATLAB function `[v,r,c]=power(A,k,tol,x)` which locates an eigenvector v for A using the power method. k is the maximum number of iterations, c is the count of the actual number of iterations, tol is the check for closeness of successive iterations, and x is the seed vector. To make your life a little easier set some default values for x=rand(n,1), tol=10^{-10}, and k=n^3. You should use the Rayleigh quotient to compute r from v.

The `nargin` feature of MATLAB allows you to set default values. See `vecplot` on page 43 of *Graphics* for an example of how to use `nargin`. Test your `power` on each of the matrices in problem 1. Let

$$A = \begin{bmatrix} 2 & 0 \\ 0 & -2 \end{bmatrix}$$

and start `power` with `x=[.8; 0]`. Now start it with `x=[0; .8]`. What happens?

3. Use the inverse shift method to find all of the eigenvalues of `h=hilb(4)` by carefully choosing the shift *s*. First try using the diagonal entries of h for the shift *s*. If you do not succeed in getting all of the eigenvalues this way, you may use `eig(h)` as a guide to choosing *s*.

4. Write a MATLAB function `[v,r,c]=invshift(A,k,tol,x,s)` which finds an eigenvector v and an eigenvalue r for A using the inverse shift method. The shift is given by `s`, while the other parameters have the same meaning as in problem 1. You can set defaults as in problem 1. Test your `invshift` on `hilb(4)`.

5. Define `A=diag(1:5)+diag(.2*ones(1,4),1)+diag(.2*ones(1,4),-1)` and plot the Gershgorin Discs and the eigenvalues. Using the inverse shift method find all of the eigenvalues by using the diagonal entries for the values of the shift, *s*.

6. Write a MATLAB plotting program `gershplot(A)` which plots the Gershgorin Discs and places the eigenvalues in them. You may use `eig`. Try your `gershplot` on `pascal(5)` and on the matrix in problem 5.

Markov Chains

ABSTRACT

An application to probability theory, Markov chains offer numerous examples. In this project we will consider some of the possibilities of iterating the transition matrix of a Markov process.

MATLAB COMMANDS

    *, sum, mesh

LINEAR ALGEBRA CONCEPTS

Eigenvalue, Eigenvector

BACKGROUND

Suppose that a particle can be in any of the states $S_1, \ldots, S_n$ and that the probability of the particle passing to state S_i from state S_j in a unit period of time is p_{ij}. Caution: Notice the order of the subscripts! We can form the matrix P whose entries are the probabilities $P(i, j) = p_{ij}$. This matrix is called a *transition matrix* or a *stochastic matrix*. A transition matrix is defined by $P(i, j) \geq 0$ and the sum of each column is 1, *i.e.* $\sum_{i=1}^{n} p_{ij} = 1$. This latter property states that the probability that the particle passes from state S_j to some other state is 1. We will begin by considering a *Markov diagram* as depicted in this figure:

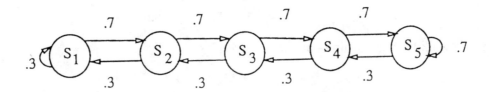

The Markov diagram pictured here represents a *random walk* since we can view it as either a move forward with probability .7 or a move backwards with probability .3. The transition matrix for this chain is

$$P_1 = \begin{bmatrix} .3 & .3 & 0 & 0 & 0 \\ .7 & 0 & .3 & 0 & 0 \\ 0 & .7 & 0 & .3 & 0 \\ 0 & 0 & .7 & 0 & .3 \\ 0 & 0 & 0 & .7 & .7 \end{bmatrix}$$

The states S_1 and S_5 form retaining barriers for the walk. This can be formed in MATLAB easily.

    v=ones(4,1);
    P1=.3*diag(v,1)+.7*diag(v,-1);
    P1(1,1)=.3; P1(5,5)=.7

189

A random walk without barriers is depicted as

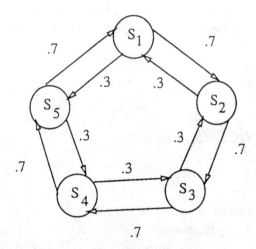

The transition matrix for this chain is

$$P_2 = \begin{bmatrix} 0 & .3 & 0 & 0 & .7 \\ .7 & 0 & .3 & 0 & 0 \\ 0 & .7 & 0 & .3 & 0 \\ 0 & 0 & .7 & 0 & .3 \\ .3 & 0 & 0 & .7 & 0 \end{bmatrix}$$

A random walk with absorbing barriers is given by

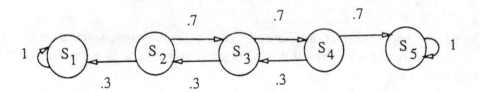

with transition matrix

$$P_3 = \begin{bmatrix} 1 & ,3 & 0 & 0 & 0 \\ 0 & 0 & .3 & 0 & 0 \\ 0 & .7 & 0 & .3 & 0 \\ 0 & 0 & .7 & 0 & 0 \\ 0 & 0 & 0 & .7 & 1 \end{bmatrix}$$

There are many other examples of Markov chains which are not random walks. We can get an example of a transition matrix straight from MATLAB

```
P=magic(5); s=sum(P); P=(1/s(1))*P
```

A *probability vector* is a vector v such that $v \geq 0$ and $\sum_{i=1}^{n} v(i) = 1$. If $v(i)$ is the probability that the particle is in state i, then $(Pv)(i) = \sum_{j=1}^{n} P(i,j)v(j)$ is the probability that the particle will be in state i in the next unit of time. It seems reasonable that multiplication of a probability vector by a transition matrix yields a probability vector. This is the content of the next theorem.

Theorem 1. *If P is a transition matrix and v is a probability vector, then Pv is a probability vector. Thus, P^n is a transition matrix for all $n \geq 0$.*

PROOF: $Pv = \sum_{j=1}^{n} v(j)P(:,j) \geq 0$. Also

$$\text{sum}(Pv) = \sum_{j=1}^{n} v(j)\text{sum}(P(:,j)) = \sum_{j=1}^{n} v(j) = 1.$$

Thus Pv is a probability vector. Let e_i be the eth standard basis vector. Now Pe_i is the ith column of P, which is a probability vector, so that $PPe_i = P^2 e_i$ is a probability vector. Since each column of P^2 is a probability vector, P^2 is a transition matrix. Similarly for P^n. ∎

We are interested in the probabilty of a move from one state to another after a long period of time. Suppose that v is a probabilty vector. The sequence $v, Pv, P^2v, P^3v, \ldots$ represents the redistribution of probabilities after successive units of time. Each $P^n v$ is called a ***state vector*** and shows the distribution of probabilities at time n. A vector s is called a ***steady state vector*** if it has the property that $Ps = s$. Notice that s is an eigenvector for the eigenvalue $\lambda = 1$. Finding the steady state vector is an interesting problem. One way, is to iterate $P^n v$ to find the dominant eigenvalue (see *The Power Method.*) Try this in MATLAB for the transition matrices given above. First let P be the magic square example. As we iterate the power we will use mesh to visually see what is happening.

```
Q=P; for i=1:35, Q=P*Q; mesh(Q), end
```

Notice how the values stabilized quickly, and for this transition matrix, each column is the same vector s=[.2 .2 .2 .2 .2]'. Now try the random walk without barriers, P_2, which we will represent in MATLAB as P2. You will need to make P2 to do this.

```
Q2=P2; for i=1:50, Q2=P2*Q2, mesh(Q2), end
```

While the convergence is slower each column again ends up as s=[.2 .2 .2 .2 .2]'. Now try the random walk with absorbing barriers, P_3, which we will represent as P3 in MATLAB.

```
Q3=P3; for i=1:50, Q3=P3*Q3, mesh(Q3), end
```

This time there is convergence, but the absorbing states S_1 and S_5 do not permit all of the columns to converge to the same steady state. Finally try the random walk with barriers

```
Q1=P1; for i=1:50, Q1=P1*Q1, mesh(Q1), end
```

The next theorems explain some of this behavior. The proof is included here since it is not normally found in a linear algebra textbook.

Theorem 2. *If P is a transition matrix, then*

    (1) $\lambda = 1$ *is an eigenvalue*

(2)   If λ is any eigenvalue, then $|\lambda| \leq 1$.
(3)   There is a probability vector v with $Pv = v$.

PROOF: 1) We know that the sum of the columns of P is 1, thus the sum of each column of $A = I_n - P$ is $1 - 1 = 0$. So the rows of A are linearly dependent and we see that the rank$(A) < n$, and thus there is a nonzero solution to $Ax = (I_n - P)x = \vec{0}$.

2) Let λ be any eigenvalue of P. By Gershgorin's Theorem (see *The Power Method*) there is an i such that

$$|\lambda - p_{ii}| \leq \sum_{j \neq i} |p_{ij}| = 1 - p_{ii}.$$

Gershgorin's Theorem has been applied to P^T, since P and P^T have the same eigenvalues. Notice that

$$|\lambda| - p_{ii} \leq |\lambda - p_{ii}| \leq 1 - p_{ii}.$$

It follows that $|\lambda| \leq 1$.

3) Let $S_n = 1/n \sum_{k=0}^{n-1} P^k$, then S_n is a transition matrix and $PS_n = S_n + (P^n - I_n)/n$. By the Bolzano-Weierstrass Theorem (this may be your first exposure to this fundamental theorem) there is a function r such that

$$\lim_{n \to \infty} S_{r(n)} = S \text{ exists}$$

Look at

$$PS = \lim_{n \to \infty} PS_{r(n)} = \lim_{n \to \infty} S_{r(n)} + \lim_{n \to \infty} (P^{r(n)} - I_n)/r(n)$$

Since $\lim_{n \to \infty} 1/r(n) = 0$ and abs$(P^{r(n)} - I_n) < $ ones(n), we get $PS = \lim_{n \to \infty} S_{r(n)} = S$. Let $v = Se_1 = \lim_{n \to \infty} S_{r(n)}e_1 \geq 0$ then

$$\text{sum}(v) = \lim_{n \to \infty} \text{sum}(S_{r(n)}e_1) = 1. \quad \blacksquare$$

Notice that for P_3 there are two steady state vectors given by the two different column vectors in the limit, e_1 and e_5. A transition matrix P is called ***regular*** if there is an n such that P^n has only positive entries. This is the case with all of the transition matrices except P_3. When $P^n > 0$ it is possible to get to any state i from any other state j, since $P^n(i, j) > 0$ for all i and all j. With a regular matrix something quite extreme happens; no matter which probability vector you start with, the sequence will tend to the same steady state probability vector.

Theorem 3. *P is regular if and only if there is a steady state probability vector $s > 0$ such that $Ps = s$ and for all probability vectors v, $\lim_{n \to \infty} P^n v = s$. Thus there is a unique steady state vector.*

PROOF: Assume P is regular. Since we know some power of P has the feature that all entries are positive, we will assume that P has this property. Let $\epsilon > 0$ be the smallest entry of P. If we choose $K = 1 - \epsilon$, then it can be shown that for probability vectors x and y

$$\|Px - Py\|_\infty \leq K\|x - y\|_\infty.$$

See *Norms and Condition Numbers* for the definition of the infinity norm which is being used here. By the Contraction Mapping Theorem (yet another theorem that you may be seeing for the first time) there is a unique fixed point $s = \lim_{n \to \infty} P^n v$, regardless of choice of v. Since $P > 0$, $s(i) = \sum_{j=1}^{n} P(i, j)s(j) > 0$ and thus $s > 0$.

Now assume that for all probability vectors v, $\lim_{n \to \infty} P^n v = s > 0$. In particular, for $v = e_i$, $\lim_{n \to \infty} P^n e_i = s$ and thus

$$\lim_{n \to \infty} P^n = S > 0 \text{ where } S = [s, \dots, s].$$

It follows that for large n, $P^n(i, j) > 0$ and P is regular. ∎

Theorem 4. *If P is regular and $Px = x$ where x is any vector, then there is a scalar r such that $x = rs$ where s is the steady state vector for P.*

PROOF: Note that $P^n x = x$ for all n and so

$$\lim_{n \to \infty} P^n x = (\lim_{n \to \infty} P^n)x = Sx = x$$

where $S = [s, \dots, s]$. Now look at $Sx = x$. $Sx = \sum_{i=1}^{n} x(i)s = rs$ where $r = \sum x(i)$ ∎

Thus finding the steady state vector for a regular matrix is just a matter of finding an eigenvector for $\lambda = 1$ and scaling. Finding probability vectors which are eigenvectors for $\lambda = 1$ when P is not regular takes more care.

PROBLEMS

1. Find the transition matrix of the following Markov diagram. What are the steady state vectors for this matrix? Is the matrix regular?

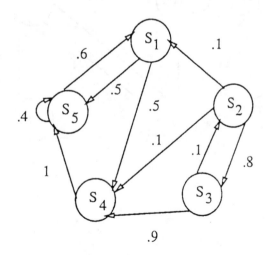

2. Find the transition matrix of the following Markov diagram. What are the steady state vectors for this matrix? Is the matrix regular?

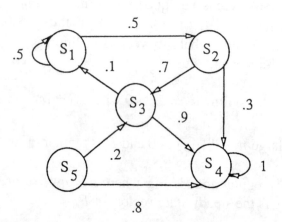

3. Since a steady state vector is an eigenvector for the eigenvalue $\lambda = 1$, we can use [B,D]=eig(P) to find the steady state vectors of P. Use this method on the matrices P_1, P_2, P_3, and the matrices in problems 1 and 2.

4. Write a MATLAB function s=steady(P) which uses MATLAB's eig to find a steady state vector. Try your steady on the matrices in problem 3.

5. Suppose that we toss a fair coin (the probability of getting H is .5) many times. We will look at the previous two tosses. There are four possibilities:

HH, HT, TH, TT

which determine four states. This gives the following Markov diagram. Find the transition matrix for this diagram. Is it regular? What is a steady state vector?

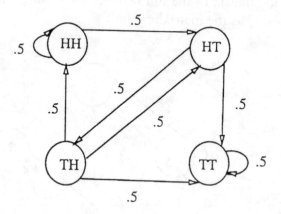

6. The Gambler's Ruin is a type of random walk. We have two gamblers, I and II. Gambler I has k dollars while gambler II has l dollars, and we let $n = k+l$. We have states $S_0, \ldots, S_n$, with gambler I in S_i if he holds i dollars. They play successive games. If gambler I wins the current game, then I wins 1 dollar and advances from S_i to S_{i+1}. Suppose that if I is in state S_i, I wins with probability p_i and II wins with probability $1 - p_i$. If I lands in state S_0 then we say I is ruined, and if I lands in state S_n, then II is ruined. Here is an example:

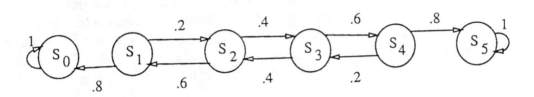

Find a transition matrix for this diagram. What is the probability that I is ruined if I begins with 3 dollars? What is the probability that II is ruined if II begins with 3 dollars? Is this matrix regular? Now increase the number of states to 11, so that they have 10 dollars to play with, while retaining the same pattern for the probabilities

$$p_{i+1,i} = \frac{i}{10} \qquad p_{i-1,i} = \frac{10-i}{10}$$

What is the probability that II is ruined if I begins with 6 dollars? 7 dollars? 8 dollars?

7. The Drop-Add line at a college has a single server and a queue which holds a maximum of n students. In a unit of time the server will handle at most one student and at most one student will attempt to join the queue. Let p be the probability that there is an arrival in a unit of time and $q = 1 - p$. Let r be the probability that the server finishes in a unit of time and $s = 1 - r$. We will let state S_i be the total number of people either in the queue or being served, thus there are $n + 1$ states, $S_0, \ldots, S_n$. The probability of a move from S_i to S_{i+1} for $0 < i < n$ is the probability that there is a new arrival and that the server is not finished, ps. The probability of a move from S_i to S_{i-1} is the probabililty that there is no new arrival and the server finishes, qr. The probability of staying in place is $t = 1 - ps - qr$. For S_0 the probability of staying in place is the probability that no one enters, q, and the probability of moving to S_1 is the probability that someone enters, p. For S_n the probability of staying in place is the probability that the server does not finish, s, while the probability of moving to S_{n-1} is the probability that the server finishes, s. We get the following diagram:

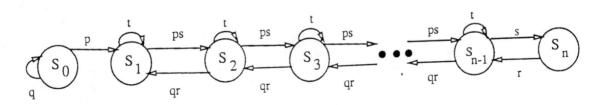

Suppose that $n = 20$, $p = .2$, and $r = .4$. Find the transition matrix for this chain. Show that the matrix is regular. Now determine the steady-state vector. The entries in the steady-state vector can be interpreted as the long range probabilities of being in a given state. What is the probability that there is no one in the queue and the server is napping? What is the probability that there is one person in the queue? What is the probability that there are less than 8 people in the queue?

8. Write a MATLAB function b=regular(P) which returns b=1 if P is regular and b=0 otherwise. You may use eig. Try your function regular on the five examples in 3.

9. The reader should be familiar with *Polynomials* before attempting this problem. It is possible to compute powers P^n quickly using the Cayley-Hamilton Theorem (see problem 15 in *Eigenvalues*). The Cayley-Hamilton says that if $f(x)$ is the characteristic polynomial of P, then $f(P) = 0$. Now if [q,r]=deconv(g,f) where $g(x) = x^m$ and if $Q(x)$ is the polynomial represented by q and $R(x)$ is the polynomial represented by r, then

$$x^m = Q(x)f(x) + R(x)$$

Thus $P^m = Q(P)f(P) + R(P) = R(P)$, since $f(P) = 0$. Now you can apply this repeatedly. Compute P^{64} in this way where P is given in problem 2.

COMMENTS

The computation in problem 9, while amusing, is not practical, since the MATLAB function poly finds characteristic polynomial by computing the eigenvalues of P first. See the comments at the end of *Eigenvalues*.

Projective Geometry

ABSTRACT
This is an introduction to the concepts of projective geometry and the transformations on projective space. Projective geometry can be applied to computer graphics.

MATLAB COMMANDS
    *, null

LINEAR ALGEBRA CONCEPTS
Null Space, Dimension, Linear Transformation

BACKGROUND
We are accustomed to using coordinate pairs for the Euclidean plane and triples for three dimensional space. In projective geometry we use triples for the plane and quadruples for three dimensional projective space. The coordinates used are called homogeneous coordinates.

Throughout this project we will use row vectors and multiplication will place the vector on the left of a matrix.

An ordered triple $[x, y, w]$ where $w \neq 0$ is the vector of **homogeneous coordinates** for the point $(\frac{x}{w}, \frac{y}{w})$. The homogeneous coordinates $[2, 4, 2]$, $[1, 2, 1]$, $[-6, -12, -6]$ all represent the point $(1, 2)$. Thus every point has infinitely many homogeneous coordinates. This is a familiar situation to those who have seen polar coordinates in Calculus. If we start with a point (x_0, y_0) we can get all the homogeneous coordinates by looking at the set of all $[wx_0, wy_0, w]$ where $w \in R$, $w \neq 0$ which in Euclidean 3-space is a line which would go through the origin, but the origin has been deleted. We will usually work with a "representative" vector of coordinates for the point (x_0, y_0) given by taking $w = 1$, $[x_0, y_0, 1]$. We will call the process of passing from $[x, y, w]$ to $[\frac{x}{w}, \frac{y}{w}, 1]$ **collapsing.** The terminology is intended to suggest that the "line" of homogeneous coordinates $\{[wx_0, wy_0, w] : w \in R, \ w \neq 0\}$ is being collapsed to the point (x_0, y_0).

When we let $w = 0$ we get a **point at infinity,** that is, $[x, y, 0]$ represents a point at infinity for each x and y, where not both x and y are 0. The collection of all these points at infinity is referred to as the **line at infinity.** The set of all collapsed homogeneous coordinates together with the line at infinity is called the **real projective plane,** denoted RP^2.

We have a similar definition of homogeneous coordinates for points in 3-dimensional space; a quadruple $[x, y, z, w]$ is the vector of homogeneous coordinates for $(\frac{x}{w}, \frac{y}{w}, \frac{z}{w})$. Collapsing is done by dividing to get the representative $[\frac{x}{w}, \frac{y}{w}, \frac{z}{w}, 1]$. Those homogeneous coordinates $[x, y, z, 0]$ represent points at infinity, and in this case the set of all points at infinity is called the **plane at infinity.** The set of homogeneous coordinates $[x, y, z, 1]$ together with the plane at infinity is the **real projective space,** written RP^3.

Advantages of This System
All Linear Equations Become Homogeneous

Look at the equation $ax + by + cz = d$. In homogeneous coordinates x is replaced by $\frac{x'}{w'}$, y by $\frac{y'}{w'}$ and z by $\frac{z'}{w'}$ yielding $\frac{ax'}{w'} + \frac{by'}{w'} + \frac{cz'}{w'} = d$. On multiplying by w we get the homogeneous equation

$$ax' + by' + cz' - dw' = 0$$

The coefficients $[a, b, c, -d]$ are called the **homogeneous coordinates for the plane.** Suppose that $H = [a, b, c, -d]$ are the homogeneous coordinates for a plane and $Q = [x, y, z, w]$ are the homogeneous coordinates for a point, then

$$Q \text{ is on the plane} \iff QH^T = 0.$$

Thus all systems of linear equations are homogeneous and hence, consistent. Notice that the plane has become a subspace, the Null Space of H, even though the plane may not pass through the origin of R^3. This is because the origin of RP^3 has homogeneous coordinates $[0, 0, 0, 1]$ and so is not the zero vector. Similarly, in RP^2 the homogeneous coordinates $[a, b, -c]$ represent the line $ax + by = c$.

Any Two Lines Intersect

Of course, two nonparallel lines will intersect. Consider the equations of two parallel lines

$$ax + by = c_1$$
$$ax + by = c_2$$

when we convert these to homogeneous coordinates

$$ax' + by' = c_1 w'$$
$$ax' + by' = c_2 w'$$

we see that $[b, -a, 0]$ is a solution, which is a point at infinity. A line will meet all lines which are parallel to it at the same point at infinity.

Translation is a Linear Transformation

In general if T is a translation transformation, $T(\vec{0}) \neq \vec{0}$, and so T will not be a linear transformation. This problem can be overcome with homogeneous coordinates. Suppose that we want to translate the origin $[0, 0, 1]$ to the point $[h, k, 1]$. Recall that matrix multiplication is a linear transformation, so we form the matrix

$$\begin{bmatrix} 1 & 0 & 0 \\ 0 & 1 & 0 \\ h & k & 1 \end{bmatrix}$$

and observe that

$$[x, y, 1] \begin{bmatrix} 1 & 0 & 0 \\ 0 & 1 & 0 \\ h & k & 1 \end{bmatrix} = [x + h, y + k, 1]$$

which are the homogeneous coordinates of the translation.

Perspective Projection is a Linear Transformation

Given a point, C, and a line, L, the perspective projection of RP^2 onto the line L is depicted in the diagram below. The point C is called the ***center of perspectivity.*** We determine the projected point $P(Q)$ by drawing a line through C and Q and locating the point of intersection with the line L. The point Q is projected onto the point $P(Q)$ on the line L. Those points Q where the line through C and Q is parallel to the line L (like Q_4) will meet L on a point at infinity, called a ***vanishing point.***

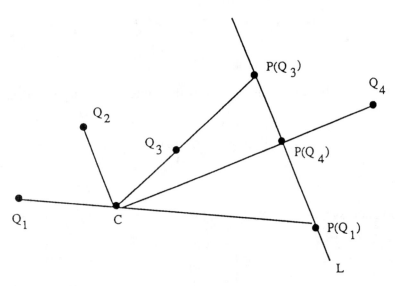

Given a center of perspectivity C and a plane, the ***perspective projection*** of RP^3 onto the plane is shown below.

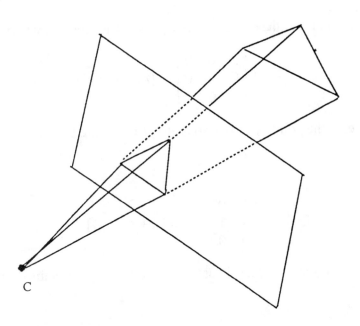

We will return to the discussion of perspectivities shortly. It is not clear at this stage that this is a linear transformation.

Transformations

Here is a short catalog of linear transformations on the projective plane. In all cases it is assumed that the coordinate vector is a row vector being multiplied on the left of the matrix.

Translation

As we have seen above the following matrix performs a translation of the homogeneous coordinates.

$$[x, y, 1] \begin{bmatrix} 1 & 0 & 0 \\ 0 & 1 & 0 \\ h & k & 1 \end{bmatrix} = [x + h, y + k, 1]$$

Rotation

The upper 2×2 submatrix of a rotation is the same as a rotation matrix in R^2.

$$\begin{bmatrix} \cos(\theta) & -\sin(\theta) & 0 \\ \sin(\theta) & \cos(\theta) & 0 \\ 0 & 0 & 1 \end{bmatrix}$$

Reflection

A reflection across the $x-$axis is given by

$$\begin{bmatrix} 1 & 0 & 0 \\ 0 & -1 & 0 \\ 0 & 0 & 1 \end{bmatrix}$$

What does a reflection matrix across the $y-$axis looks like?

Scaling

The following will mutiply the x coordinate by r and the y coordinate by s.

$$[x, y, 1] \begin{bmatrix} r & 0 & 0 \\ 0 & s & 0 \\ 0 & 0 & t \end{bmatrix} = [rx, sy, t]$$

The outcome represents the point $(\frac{r}{t}x, \frac{s}{t}y)$, so that $1/t$ is scaling both x and y.

Shear

A shear is given by

$$[x, y, 1] \begin{bmatrix} 1 & 0 & 0 \\ s & 1 & 0 \\ 0 & 0 & 1 \end{bmatrix} = [x + sy, y, 1]$$

This is a shear in the x direction. The shear in the y direction is obtained by placing s in the $(1, 2)$ position.

There are three dimensional analogues of these transformations which we will take up in *Computer Graphics*

Subspaces

Let W be a subspace of $\mathrm{R}^4 = \{[x, y, z, w] : x, y, z, w \in \mathrm{R}\}$. By collapsing the homogeneous coordinates in W to $[\frac{x}{w}, \frac{y}{w}, \frac{z}{w}, 1]$, we get an object of RP^3, called a ***projective subspace***. A projective subspace is either a point (if $\dim(W) = 1$), a line (if $\dim(W) = 2$), a plane (if $\dim(W) = 3$), or all of RP^3 (if $\dim(W) = 4$). As we have seen, the equations which define the projective subspaces can be turned into homogeneous equations using homogeneous coordinates, and so each projective subspace can be viewed as a null space of some matrix.

Suppose that we have two points in RP^2 with homogeneous coordinates given by $P_1 = [a_1, b_1, 1]$ and $P_2 = [a_2, b_2, 1]$, then the subspaces $W_1 = \mathrm{span}(P_1)$ and $W_2 = \mathrm{span}(P_2)$ are 1-dimensional subspaces. Assuming $P_1 \neq P_2$, the line through P_1 and P_2 is a projective subspace given by the 2-dimensional subspace $W_1 + W_2 = \mathrm{span}(P_1, P_2)=$Row Space of $[P_1; P_2]$. If we want to represent this subspace as a null space, consider the line in RP^2 with homogeneous coordinates $L = [b_1 - b_2, a_2 - a_1, a_1 b_2 - a_2 b_1]$. It is easy to check that $P_1 \cdot L = 0$ and $P_2 \cdot L = 0$. The last coordinate $a_1 b_2 - a_2 b_1$ is the determinant of $[P_1; P_2]$.

Perspectivities

In RP^2 we can perspectively project onto a line. In RP^3 we can perspectively project onto a plane. The matrix for doing this has a simple definition.

Theorem 1. *Let H be the homogeneous coordinates for a line in RP^2 (or a plane in RP^3) and let C be the homogeneous coordinates of a point in RP^2 (resp. a point in RP^3.) The perspectivity with center C onto the line (resp. plane in RP^3) is a linear transformation given by the matrix*

$$P = (HC^T)I_n - H^T C$$

where $n = 3$ (resp. $n = 4$.)

PROOF: First we show that for any point Q the point QP is on the line, by showing that $(QP)H^T = 0$.

$$QPH^T = Q(HC^T)I_n H^T - QH^T(CH^T)$$
$$= (HC^T)(QH^T - QH^T) = 0$$

since $CH^T = HC^T$.

Now we need to show that QP is on the line through C and Q. Let A be a matrix where the Null Space of A is the projective space for the line through Q and C. We will show $(QP)A = \vec{0}$. Note that $QA = CA = \vec{0}$, thus

$$(QP)A = QHC^T A - QH^T CA = HC^T(QA) - QH^T(CA) = \vec{0}$$

since HC^T is a scalar. ∎

Example of a Perspective Projection

We will look at some perspective projections. We will place a triangle in the plane $z = 1$ and project it onto the $xy-$plane. We will move the center of the perspectivity around to get different views of the triangle. First create the triangle and plot it to see if it looks right.

```
t=[0;2*pi/3;4*pi/3;0];
x=cos(t); y=sin(t); plot(x,y)
d=[x,y,ones(t),ones(t)];
```

This glues the triangle into one matrix. The third column puts it on the $xy-$plane and the fourth column gives it homogeneous coordinates. The $xy-$plane has the equation $z = 0$, so the homogeneous coordinates for the plane are [0, 0, 1, 0]. Next pick a center of perspectivity. Let the first center be $(0, 0, 2)$ which has homogeneous coordinates $C = [0, 0, 2, 1]$. Now we will use Theorem 1, to find the matrix of the perspectivity.

```
P=[0,0,1,0]*[0,0,2,1]'*eye(4)
P=P-[0,0,1,0]'*[0,0,2,1]
newd=d*P
```

This gives the homogeneous coordinates for the projected triangle. We may have entries other than 1 in the fourth column. To rectify this we collapse by dividing each row by the entry in the fourth column.

```
w=newd(:,4)*ones(1,4), newd=newd./w,
```

To plot we need two vectors.

```
x=newd(:,1); y=newd(:,2);
```

Before plotting set the axis with `axis([-5,5,-5,5])`. Now do

```
plot(x,y)
```

Put a hold on that plot and try some other centers like $(0, 0, 10)$, $(4, 0, 2)$, $(4, 0, -1)$. You will have to recreate the perspective matrix P for each new center.

PROBLEMS

1. Make the following data matrices

```
Gx=[0 .2 2 2.2 2.2 6 6.3 6 2.2 0 -.2 0 0]'
Gy=[0 2 2 0 2.5 5.5 5.5 6 3 3 3.1 2.9 0]'
```

Try `plot(Gx,Gy)` to see what it is. Now set the axis at

```
axis([-12,12,-12,12])
```

Using homogeneous coordinates and the transformation matrices do the following:

    (1)   reflect around the $x-$axis
    (2)   rotate by 30 degrees
    (3)   translate to (1,2)
    (4)   scale the entire figure by a factor of 1.5
    (5)   shear by a factor of 1.5 in the $y-$direction.

2. The classic situation in projective geometry occurs when two parallel lines are viewed in perspective. Make the following data matrices

```
Ly = [1 1 2 2]'
```

`Lz=[0 200 200 0]'` Run `plot(Ly,Lz)` to see what you have. We are first going to project into the $xy-$plane. Place the lines in the $yz-$plane and give them homogeneous coordinates.

```
d=[zeros(4,1),Ly,Lz,ones(4,1)]
```

Use `c=[2, 1.5, -1, 1]` for the center of perspectivity. You will notice that the lines will trail-off to a point, the *vanishing point.* Move the center to `c=[1, 1.5, -1, 1]` and notice how the vanishing point moves with it.

3. Write a MATLAB function `[x,y]=persproj(c,d)` which finds the x and y coordinates of the perspective projection of the object with $m \times 3$ data matrix d of points (the rows of d) onto the $xy-$plane with center of perspectivity c. Be sure to divide by the last coordinate of the homogeneous coordinates to prepare for plotting as we did in the example.

4. Try out your `persproj` on the example in the Background discussion.

5. Make a 2-dimensional graphics library consisting of the following functions:

In each of these the $m \times 2$ matrix d represents an object given by rows of d.

    (1)   `[x,y]=rotate2(t,d)` rotates the object through an angle of t radians.

    (2)   `[x,y]=refx(d)` and `[x,y]=refy(d)` which reflect the object around the x and the y axes respectively.

    (3)   `[x,y]=trans2(h,k,d)` which translates the object by translating $(0,0)$ to (h,k).

    (4)   `[x,y]=shear2(r,s,d)` which shears the object by r in the x direction and by s in the y direction.

    (5)   `[x,y]=scale2(r,s,t,d)` which scales the object by r in the x coordinate, s in the y coordinate, and by t in both coordinates.

In these functions you should collapse `[x,y,w]` to `(x/w,y/w,1)`.

6. Try your graphics library on the giraffe in problem 1. Create an animation sequence where the giraffe leaps over an imaginary hurdle.

7. Set up the triangle in the plane with vertices at $(0,0)$, $(2,0)$, and $(1,1)$ using homogeneous coordinates. Now use the functions in your graphics library to rotate the triangle through an angle of $\pi/3$ around the point $(3,2)$. Note the usual rotation rotates around $(0,0)$.

8. Set up a circle in the $yz-$plane using

```
t=0:pi/25:2*pi; y=cos(t); z=sin(t);
```

Project the circle onto the $xy-$plane using `persproj` by setting the center of projectivity at

```
c=[100,0,-1,1]
```

You will get a message that infinity has been found in the data. When you plot the result, keep in mind that MATLAB has plotted across the non-infinite data. Thus the parabola you see, with the ends connected is really a parabola. Try to determine where the vanishing point is occurring. Try this again with the center at $(100, 0, 1)$. Now move the line for the center of perspectivity up to $(100, 0, -.5)$ and try the same experiment. This time there are two vanishing points and the result is a hyperbola. Again MATLAB plots across the points at infinity. Finally move the center to $(100, 0, -2)$. This time there are no vanishing points and the result is an ellipse. In projective geometry a conic can be projected into a circle. What positions for C produce a parabola? an ellipse? a hyperbola?

9. The following data matrix will set up three faces of a cube in three dimensional space:

$$d = \begin{bmatrix} 0 & 0 & -1 \\ 0 & 1 & -1 \\ 1 & 1 & -1 \\ 1 & 0 & -1 \\ 0 & 0 & -1 \\ 0 & 1 & -1 \\ 0 & 1 & -2 \\ 0 & 0 & -2 \\ 0 & 0 & -1 \\ 1 & 0 & -1 \\ 1 & 0 & -2 \\ 0 & 0 & -2 \end{bmatrix}$$

Using persproj project it onto the xy-plane using C=[-1,-1,1,1]. Plot the result setting the axis to v=[-2,2,-2,2]. Now move the center of perspectivity back to C=[-5,-5,5,1], to C=[-10,-10,10,1]. Notice how some of the lines tend to move towards each other, so that if you were to extend them they would meet at a vanishing point at infinity. Now move to infinity C=[-1,-1,1,0]. Here the lines are parallel. When the center of perspectivity is placed at infinity we get a ***parallel projection.***

10. The following data matrix will trace a tetrahedron sitting on the plane $z = 1$ in homogeneous coordinates.

$$d = \begin{bmatrix} 1 & 0 & 1 & 1 \\ \cos(2\pi/3) & \sin(2\pi/3) & 1 & 1 \\ \cos(4\pi/3) & \sin(4\pi/3) & 1 & 1 \\ 1 & 0 & 1 & 1 \\ 0 & 0 & 2 & 1 \\ \cos(4\pi/3) & \sin(4\pi/3) & 1 & 1 \\ \cos(2\pi/3) & \sin(2\pi/3) & 1 & 1 \\ 0 & 0 & 2 & 1 \end{bmatrix}$$

Project this tetrahedron onto the xy-plane using a center at $(x, 0, -1)$. Now try letting $x = 0, 1, 2, 3, 4, 5$. When you plot place an 'o' on the top vertex by holding the plot and calling plot(x(5),y(5),'o'). This will help your orientation in viewing.

Now fix the center at $(0, 0, -1)$ and rotate the tetrahedron in increments of $\pi/10$ from 0 to $\pi/2$ around the x-axis using a rotation matrix

$$\begin{bmatrix} 1 & 0 & 0 & 0 \\ 0 & \cos(t) & \sin(t) & 0 \\ 0 & -\sin(t) & \cos(t) & 0 \\ 0 & 0 & 0 & 1 \end{bmatrix}$$

You may want to mark the vertex again.

COMMENTS

We have only offered very primitive projections into the xy−plane in this project. A viewing system where we can view a projection onto any plane is set up in *Computer Graphics*.

Computer Graphics

ABSTRACT

We develop a small library for interactive viewing and manipulation of three dimensional objects. We will be able to view the perspective projection onto any plane. *Projective Geometry* is a prerequisite for this project.

MATLAB COMMANDS

*, ', \

LINEAR ALGEBRA CONCEPTS

Perspective Projection, Rotation, Translation, Scale

BACKGROUND

Projections

Let us begin by making the following prism in MATLAB:

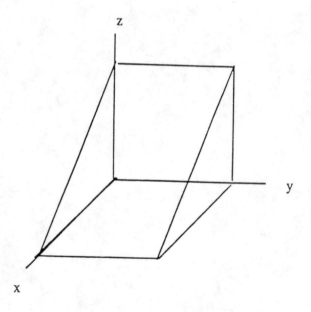

This can be done in homogeneous coordinates (see page 197 in *Projective Geometry*) with the data matrix:

$$\text{prism} = \begin{bmatrix} 0 & 0 & 0 & 1 \\ 1 & 0 & 0 & 1 \\ 0 & 0 & 1 & 1 \\ 0 & 0 & 0 & 1 \\ 0 & 1 & 0 & 1 \\ 1 & 1 & 0 & 1 \\ 1 & 0 & 0 & 1 \\ 0 & 0 & 1 & 1 \\ 0 & 1 & 1 & 1 \\ 0 & 1 & 0 & 1 \\ 1 & 1 & 0 & 1 \\ 0 & 1 & 1 & 1 \end{bmatrix}$$

You should verify by hand that plotting this data does produce the picture in 3-dimensions.

We are going to make a plot of prism in MATLAB but we need to project it onto a plane first. In *Projective Geometry* we did some simple projections onto coordinate planes, but those will give an uninteresting view of prism. A good vantage would be in the direction of the vector $[1, 1, 1]$. A projection onto the plane $x + y + z = d$ from a point at infinity, $[1, 1, 1, 0]$ is called an **isometric projection.** We are going to make an isometric projection of prism.

Recall from Theorem 1 of *Projective Geometry* that the projection matrix is

$$P = HC^T I_4 - H^T C$$

where $H = [1, 1, 1, -d]$ is the vector of homogeneous coordinates for the plane and $C = [1, 1, 1, 0]$ is the vector of homogeneous coordinates for the center of perspectivity. Make P in MATLAB using $d = 2$.

```
H=[1,1,1,-2]; C=[1,1,1,0];
P=H*C'*eye(4)-H'*C
```

Now multiply

```
pprism=prism*P
```

and collapse the homogeneous coordinates.

```
pprism=pprism./(pprism(:,4)*ones(1,4))
pprism=pprism(:,1:3),
```

Great! Now we have the projected points but we cannot plot them since they have three coordinates. What we need is a coordinatization of the plane, a two dimensional object, so that we can represent the projected points in two coordinates.

First, we will translate them back to a plane which is parallel to $x + y + z = 2$ but passes through the origin, namely $x + y + z = 0$. This is a *subspace* of R^3, which we will coordinatize by finding an orthonormal basis.

In the diagram below the two lines represent the planes and the point Q is any point on the original plane. We want to solve for the vector v in terms of the normal vector n and Q.

We know one normal vector for the plane

```
n=H(1:3)
```

We need a normal vector which touches the plane. This will be a scalar multiple rn of the normal n. So we solve $rnH(1:3)^T = d$ which in this case gives $r = 2/3$. Let

```
n=(2/3)*n
```

We can solve $n + v = Q$ for v to get $v = Q - n$. In MATLAB we will do this all at once using an outer product (see page 32.)

```
ones(12,1)*n
```

This shows us the matrix to subtract.

```
pprism=pprism-ones(12,1)*n
```

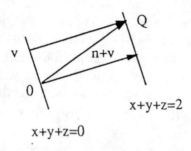

Now we are going to get an orthonormal basis for the plane. This can be done in many ways, but we will choose a basis so that the orientation is reasonable when it is time to view. This is done by choosing a vector which indicates the upward direction of the plane. For the isometric view that we are working on for this example, that vector is $v = (0, 0, 1)$. But this is not a vector in the plane. We will first make a cross product $u_1 = v \times n$, which will give us a vector in the plane, since $u_1 \cdot v = 0$, but since $u_1 \cdot v = 0$, u_1 will make a better horizontal direction. Then let $u_2 = n \cdot u_1$, which is again in the plane and this will be our vertical direction. The cross product (see page 136 in *Linear Transformations*), $v \times w$ is

$$(v(2)w(3) - w(2)v(3), v(3)w(1) - w(3)v(1), v(1)w(2) - w(1)v(2))$$

I suggest that you program this for convenient use later.

In MATLAB

```
u1=[v(2)n(3)-n(2)v(3),v(3)n(1)-n(3)v(1),
v(1)n(2)-n(1)v(2)]
u1=(1/norm(u1))*u1
```

This gives us a unit vector on the plane.

```
u2=[n(2)u1(3)-u1(2)n(3),n(3)u1(1)-u1(3)n(1),
n(1)u1(2)-u1(1)n(2)]
u2=(1/norm(u2))*u2
```

Success is now within our grasp. We need only get the coordinates for the projected points and plot them.

```
x=pprism*u1'
y=pprism*u2'
plot(x,y)
```

This may not be the orientation that you want on the object. This can be corrected by changing the sign on either x or y or both. This reflects the projected points across one of the axes. Changing the sign on either of u1 or u2 has the same effect.

Returning to the general projection problem, it is a bit cumbersome to specify the homogeneous coordinates of a plane. An easier way is to use spherical coordinates to describe the normal vector n. Recall spherical coordiates (r, θ, ϕ) for a point (x, y, z) as depicted below.

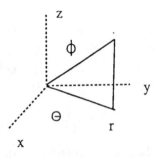

The rectangular coordinates can be recovered by the formulas

$$x = r \ \sin(\phi)\cos(\theta)$$
$$y = r \ \sin(\phi)\sin(\theta)$$
$$z = r \ \cos(\phi)$$

Notice that $\phi = 0$ specifies a point directly overhead of the $xy-$ plane on the $z-$ axis.

Three Dimensional Graphics Transformations

Throughout this discussion, points will be represented by row vectors and multiplication will be from the left, *viz.* xA.

Translation

This matrix is completely analogous to the 2-dimensional translation.

$$\text{trans}(h, k, l) = \begin{bmatrix} 1 & 0 & 0 & 0 \\ 0 & 1 & 0 & 0 \\ 0 & 0 & 1 & 0 \\ h & k & l & 1 \end{bmatrix}$$

Rotation

There are rotations around each of the coordinate axes. They are described by the Givens rotations (see *Building Matrices*).

Rotation around the x-axis

$$\text{rotx}(\theta) = \begin{bmatrix} 1 & 0 & 0 & 0 \\ 0 & \cos(\theta) & -\sin(\theta) & 0 \\ 0 & \sin(\theta) & \cos(\theta) & 0 \\ 0 & 0 & 0 & 1 \end{bmatrix}$$

Rotation around the y-axis

$$\text{roty}(\theta) = \begin{bmatrix} \cos(\theta) & 0 & -\sin(\theta) & 0 \\ 0 & 1 & 0 & 0 \\ \sin(\theta) & 0 & \cos(\theta) & 0 \\ 0 & 0 & 0 & 1 \end{bmatrix}$$

Rotation around the z-axis

$$\text{rotz}(\theta) = \begin{bmatrix} \cos(\theta) & -\sin(\theta) & 0 & 0 \\ \sin(\theta) & \cos(\theta) & 0 & 0 \\ 0 & 0 & 1 & 0 \\ 0 & 0 & 0 & 1 \end{bmatrix}$$

Scale

The following matrix will mutiply the x-coordinate by r, the y-coordinate by s, and the z-coordinate by t.

$$\text{scale}(r, s, t) = \begin{bmatrix} r & 0 & 0 & 0 \\ 0 & s & 0 & 0 \\ 0 & 0 & t & 0 \\ 0 & 0 & 0 & 1 \end{bmatrix}$$

If you want global scaling, then the $(4, 4)$ position can be used also.

Shear

A shear is given by a $\text{shear}(s, i, j) = \text{ele3}(4, s, i, j)$ where $i, j \le 3$. For example,

$$\text{shear}(s, 1, 2) = \begin{bmatrix} 1 & 0 & 0 & 0 \\ s & 1 & 0 & 0 \\ 0 & 0 & 1 & 0 \\ 0 & 0 & 0 & 1 \end{bmatrix}$$

PROBLEMS

1. Write a MATLAB function `w=veccross(u,v)` which computes the cross product of the vectors u and v.

2. Make the section of a sphere in the first octant as follows:

```
t=0:pi/80:pi/2; t=t'; x=cos(t); y=sin(t);
```

```
sphere=[x,y,zeros(t),ones(t);

zeros(t),x,y,ones(t);

y,zeros(t),x,ones(t)];
```

Get an isometric view of the sphere. If you do not understand why the coordinates are arranged in this way, try it the way that you think they should be.

3. Write a MATLAB function `[x,y]=iso(data,d,up,s,t)` which makes an isometric projection of the object given by `data` onto the plane $x + y + z = d$. The variable up is to be the upward pointing vector, you should have a default value of `up=[0,0,1]`. The variables s and t are to be ± 1 to give an orientation to the basis vectors u1 and u2 by multiplying `s*u1` and `t*u2`. They should have default values of `s=t=1`.

4. Test your `iso` on `prism` and `sphere`.

5. Define a data matrix cube which consists of three faces of a cube on the vertices $(1, 0, 0)$, $(1, 1, 0)$, $(1, 1, 1)$, $(1, 0, 1)$, $(0, 0, 1)$, $(0, 1, 1)$, $(0, 1, 0)$. Get an isometric view of cube.

6. Write MATLAB functions

    (1) `A=rotx(theta)`
    (2) `A=roty(theta)`
    (3) `A=rotz(theta)`
    (4) `A=scale(r,s,t)`
    (5) `A=shear(s,i,j)`

The definitions of these functions are given in the Background discussion.

7. Try each function in problem 5 out on `prism` and `cube` by giving rotation through angles of $pi/4$ and $pi/2$. Use `shear` on `prism` to make the front triangular face vertical.

8. Rotate cube so that the diagonal is on the $z-$axis. Now rotate it through an angle of $\pi/2$ around the $z-$ axis. Then rotate it back to its original position. Putting this all together it should tell you how to rotate cube around its diagonal.

9. Write a MATLAB function `[x,y]=newpersproj(d,plane,center,up,s,t)` which makes a perspective projection onto the plane whose normal vector is given by spherical coordinates in `plane` with center of projectivity `center`. The variables up, s and t are the same as in problem 2, and they should be given default values as in problem 2. The variable `center` should be given a default value of the point at infinity so that the default is a parallel projection.

10. Using `newpersproj` make projections of cube onto the planes

    (1) $x = 2$ with `center=[4,0,0,1]`
    (2) $x + y = 4$ with `center=[8,8,0,1]`
    (3) $x + y + z = 2$ with `center=[4,4,4,1]`

The first is a one point perspective with a single vanishing point in the direction of the negative $z-$ axis given by the point at infinity where the lines converge. The second is a two point perspective with two vansishing points in the directions of the negative $z-$ axis and the negative $y-$ axis. The third is a three point perspective with three vanishing points.

11. We can make a 3-dimensional roller coaster as follows. We will give it a circular base.

```
t=0:pi/30:2*pi;  t=t';   x=cos(t); y=sin(t);

z=cos(2*t)+cos(3*t)+sin(4*t); coast=[x,y,z,ones(t)];
```

Now use `newpersproj` to get several views of `coast`.

12. For a challenging programming project take a ride on the roller coaster in problem 10. The idea is to project onto the normal plane at each point on the curve. You will need to find the normal by taking the derivatives with respect to t. Initially you might find the picture a bit confusing, since the entire roller coaster is being projected onto the plane. The way to avoid some of this is to only project that piece of `coast` which is "ahead" of the plane. This is called clipping. It requires determining which part of the curve is ahead of the plane and removing the part which is behind. The easiest way to establish when points P and Q are on the same side of the plane with homogeneous coordinates H is by checking that $P \cdot H$ and $Q \cdot H$ have the same sign.

COMMENTS

We have not addressed two important issues for computer graphics: hidden lines and clipping. When you use `mesh`, you should have noticed that points and lines which are behind the nearest surface in our line of sight are not visible. This is because `mesh` incorporates a hidden line algorithm which suppresses those lines. Clipping allows the viewer to specify a box in 3-dimensional space and only see those objects which are inside the box.

Symmetric Diagonalization

ABSTRACT

It is always possible to diagonalize a real symmetric matrix. We will develop background on complex matrices and the Schur Decomposition to achieve this.

MATLAB COMMANDS

```
schur, eig, rsf2csf, plot
```

LINEAR ALGEBRA CONCEPTS

Symmetric Matrix, Orthogonal Matrix, Unitary Matrix, Hermitian Matrix, Positive Definite Matrix

BACKGROUND

Since the eigenvalues of real matrices can be complex numbers we are led naturally to consider complex matrices. Suppose that A is a matrix with complex entries. The MATLAB operation A' is the transpose of the conjugate (see *Complex Numbers*). If the entries of A are real, this is the same as the transpose. The MATLAB function conj(A) finds the conjugates of the entries without doing the transpose. It is common in textbooks to see the conjugate of the transpose written as A^H. The H is in honor of Charles Hermite, a developer of the following theory.

If we are given two column vectors u and v with *complex entries,* we define the dot product as

$$u \cdot v = \sum_{i=1}^{n} \overline{u_i} v_i$$

where $\overline{u_i}$ is the complex conjugate. This will share most of the properties of the dot product on vectors with real entries. The main exception is that while $u \cdot v = v \cdot u$ for real vectors, for complex vectors we have

$$u \cdot v = \overline{v \cdot u}$$

so that usually, $u \cdot v \neq v \cdot u$. We go on to define the norm on complex vectors as

$$\|v\| = \sqrt{v \cdot v} = \sqrt{\sum_{i=1}^{n} \overline{v_i} v_i}.$$

We define two nonzero vectors u and v to be **orthogonal** if $u \cdot v = 0$, and u is a **unit vector** if $\|u\| = 1$. Vectors $u_1, \ldots, u_n$ are **orthonormal** if

$$u_i \cdot u_j = \begin{cases} 0 & \text{if } i \neq j \\ 1 & \text{if } i = j \end{cases}$$

This development now parallels that of of real vectors. We call an $n \times n$ matrix **unitary** if the columns of U are orthonormal vectors, equivalently, $U^H U = I_n$. A unitary matrix is the complex version of an orthogonal matrix. We are going to consider the problem of diagonalization again. The first result shows that we can bring a matrix into triangular form using a unitary matrix.

Theorem 1. *The Schur Decomposition. Given an $n \times n$ matrix A there is a unitary matrix U and an upper triangular matrix T such that*

$$U^H AU = T$$

The eigenvalues are found in this decomposition since T and A, as similar matrices, will have the same eigenvalues and the eigenvalues of T are on the diagonal. MATLAB has a function [U,T]=schur(A) which computes the Schur Decomposition. Try this on some matrices

```
A=[1,3;2,1]; [U,T]=schur(A)
```

```
A=[1,i;1,2]; [U,T]=schur(A)
```

As you can see, it works fine. Now try

```
A=[0,1;-1,0], [U,T]=schur(A)
```

You notice that schur did not do much with this last matrix. The reason is that there are actually two Schur Decompositions, the ***real schur form*** and the ***complex schur form.*** The form described in Theorem 1 is the complex schur form. The real schur form may not produce an upper triangular matrix for T; there may be 2×2 blocks occurring on the diagonal of T. What is given up by the real schur form in the triangular matrix is regained in the fact that T and U have real entries. Try this out in MATLAB

```
A=rand(7); [U,T]=schur(A)
```

Notice the 2×2 blocks on the diagonal. Try

```
A=compan(ones(1,5)); [U,T]=schur(A)
```

The matrix compan is described in *Building Matrices* . Here again you will notice the little 2×2 blocks. Now try

```
eig(A), eig(T(1;2,1:2)), eig(T(3:4,3:4))
```

You see that the conjugate pairs of eigenvalues are falling inside the 2×2 blocks. The function schur will automatically find the real schur form of a real matrix and the complex schur form of a complex matrix. MATLAB has a function [U,T]=rsf2csf(U,T) which will convert the real schur form to the complex schur form.

A matrix A is ***Hermitian*** if $A = A^H$. If A has only real entries, then A is Hermitian if and only if A is symmetric. If A has complex entries, then Hermitian is not the same as symmetric. Consider for example

$$A = \begin{bmatrix} 1 & i \\ -i & 2 \end{bmatrix}$$

which is Hermitian but not symmetric. Hermitian matrices have real eigenvalues.

Theorem 2. *If A is Hermitian, then*
(1) A has real eigenvalues.
(2) If v_1 and v_2 are eigenvectors for distinct eigenvalues, then v_1 and v_2 are orthogonal.

This gives the Spectral Theorem.

Theorem 3. *The Spectral Theorem. If A is Hermitian, then there is a diagonal matrix D with real entries and a unitary matrix U such that*

$$U^H AU = D$$

If A is a real symmetric matrix, then the eigenvectors will be real, and U can be taken to be a real orthogonal matrix.

PROOF: Let $U^H AU = T$ be the Schur Decomposition of A. Note that T is Hermitian, since $T^H = U^H A^H U = U^H AU = T$, and since T is triangular, it must be diagonal. ∎

The function $\texttt{schur}$ will find D if A is Hermitian.

```
B=i*triu(rand(4)); A=B'+B; A-A',
[U,T]=schur(A)
```

Quadratic Forms

A **quadratic form** is a product $x^T Ax$ where A is a real symmetic matrix. For a 2×2 matrix, $[x \; y] \begin{bmatrix} a & b \\ b & c \end{bmatrix} \begin{bmatrix} x \\ y \end{bmatrix} = ax^2 + 2bxy + cy^2$. In general,

$$x^T Ax = \sum_i a_{i,i} x_i^2 + \sum_{i<j} 2a_{i,j} x_i x_j$$

By the Spectral Theorem the coefficient matrix A can be diagonalized with real matrices, $U^T AU = D$, where U is orthogonal and D is a real diagonal matrix. If we let $z = U^T x$ be an orthonormal change of variables, then

$$x^T Ax = z^T Dz = \sum_i \lambda_i z_i^2$$

Suppose we have a conic in two dimensions. These can be written in the form

$$ax_1^2 + bx_1 x_2 + cx_2^2 + dx_1 + ex_2 + f = 0$$

Our main interest is handling the $x_1 x_2$ term. This term represents a rotation of the conic and we will use the Spectral Theorem to find the rotation matrix. As an example take the conic $x^2 + 2xy - y^2 = 1$. The quadratic form is

$$[x_1 \quad x_2] \begin{bmatrix} 1 & 1 \\ 1 & -2 \end{bmatrix} \begin{bmatrix} x_1 \\ x_2 \end{bmatrix}$$

In MATLAB try

```
A=[1, 1;1, -2];
```

Since this is a real symmetric matrix we can use `schur` to diagonalize it.

```
[U,D]=schur(A)
```

This gives `U'*A*U=D` and so we can substitute for `A` into the quadratic form to get `x'*U*D*U'*x=1` and let `z=U'*x`. Thus

$$z\text{'}Dz = \lambda_1 z_1^2 + \lambda_2 z_2^2 = 1.$$

Since $\lambda_1 > 0$ and $\lambda_2 < 0$ this is a hyperbola. We can solve for $z_1 = \pm\sqrt{\frac{1-\lambda_2 z_2^2}{\lambda_1}}$. In MATLAB

```
z2=-5:.1:5; z1=sqrt((1-D(2,2)*z2)./D(1,1));
plot(z1,z2)
```

To get the other half of the hyperbola

```
z3=-z1; plot(z1,z2,z3,z2)
```

Now rotate the conic back with

```
x=U*[z1;z2]; y=U*[z3:z2];
```

and plot

```
plot(x(1,:),x(2,:),y(1,:),y(2,:))
```

The other conics will require modified plotting techniques. See problem 7 of *Linear Transformations* for plotting an ellipse.

A real symmetric matrix A is ***positive definite*** if the quadratic form $x^T A x > 0$ for all real vectors $x \neq \vec{0}$. We claim that the positive definite matrices are the symmetric matrices with positive eigenvalues. While this definition looks like it depends on all vectors x, it really only depends on the vectors x where $\|x\| = \sqrt{x_1^2 + x_2^2 + \cdots + x_n^2} = 1$. To see this, normalize a vector $y \neq \vec{0}$ to get $y = rx$ for some $r \neq 0$ and $y^T A y = (rx)^T A(rx) = (r^2)x^T ax$; so the sign is determined by x. If $U^T A U = D$, we see that being positive definite is the same as $z^T Dz > 0$ for all vectors $z = Ux$ where $x \neq \vec{0}$. Since U is invertible, $z = \vec{0} \iff x = \vec{0}$. Thus A is positive definite if and only if $z^T Dz > 0$ for all $z \neq \vec{0}$. By choosing $z = e_i$, the standard basis vector, we see that $e_i D e_i = \lambda_i$ so that A is positive definite if and only if the eigenvalues of A are positive. It also follows that $det(A) > 0$ and A is invertible. We now consider how this might be used.

Suppose that we are given a function of two variables $f(x, y)$ with the mission of finding the relative maxima and minima. We will show how this can be viewed as a quadratic form problem. Suppose that you know a critical point (z_1, z_2), having found $\frac{\partial f}{\partial x}(z_1, z_2) = 0$ and $\frac{\partial f}{\partial y}(z_1, z_2) = 0$. By Taylor's Theorem it can be shown that

$$f(x, y) \approx f(z_1, z_2) + u^T A u$$

where

$$A = \begin{bmatrix} a & b \\ b & c \end{bmatrix}$$

$$a = \frac{1}{2}\frac{\partial^2 f}{\partial x^2}(z_1, z_2)$$

$$b = \frac{1}{2}\frac{\partial^2 f}{\partial x \partial y}(z_1, z_2)$$

$$c = \frac{1}{2}(y - z_2)^2\frac{\partial^2 f}{\partial y^2}(z_1, z_2)$$

$$u = \begin{bmatrix} x - z_1 \\ x - z_2 \end{bmatrix}$$

Now if $u^T A u > 0$ for all $u \neq \vec{0}$, then (z_1, z_2) is a relative minimum and if $u^T A u < 0$ for all $u \neq \vec{0}$, then (z_1, z_2) is a relative maximum. When both positive and negative values can be attained, then we have a saddle point. When $u^T A u = 0$, the higher degree terms are significant and we cannot use quadratic form methods on this problem. Thus we have shown that if A is positive definite, there is a relative minmium and if A is **negative definite** ($x^T A x < 0$ for all $x \neq \vec{0}$, *i.e.* the eigenvalues are negative), then there is a relative maximum. If A has both positive and negative eigenvalues, then there is a saddlepoint.

PROBLEMS

1. Find the real schur form of the following: `rand(5,1)*rand(1,5)`

```
list(5)
pascal(5)
hilb(7)
rand(7)
jordan(2,5)'.
```

2. In the examples of problem 1 use the function `rsf2csf` to find the complex schur form. Try `schur` on the matrices `rand(5)+i*rand(5)` and `A=B*B'`.

3. Rotate the axis to eliminate the xy term in each of the following conics and graph in MATLAB. Then plot the original conic.

(1) $x^2 + 2xy + 2y^2 = 1$
(2) $x^2 + 12xy + 2y^2 = 1$
(3) $4x^2 - 12xy + 9y^2 = 1$

4. For each of the following determine if the point given is a relative maximum, a relative minimum, or a saddle point.

(1) $f(x, y) = \cos(x) + \cos(y)$ at $x = y = 0$
(2) $f(x, y) = xy - x^2 - 2y$ at $x = 2, y = 4$
(3) $f(x, y) = 2x^4 + y^4 - x^2 - 2y^2$ at $x = 0, y = 0$ and $x = 0, y = 1$

For each make a mesh graph to see what is happening at the point in question.

5. We have relied on matrices of the form $A = B^T B$ for examples in this project. When B is invertible, such an A is positive definite. Show this first for B=rand(5), and then prove it in general.

6. The general conic can be written $x^T Ax + Bx + C = 1$ where A is symmetric $n \times n$, B is a $1 \times n$ row vector, and C is a scalar. By applying the Spectral Theorem, $A = UDU^T$ where U is orthogonal and D is diagonal, and letting $z = U^T x$ and $E = BU$ we get $z^T Dz + Ez + C = 1$. Now you can complete the square (see your Calculus book) to translate the conic to the origin. In this form it is easily identified and plotted. You can then translate it back to the original center and rotate it into correct position. Try this on

$$7x^2 - 12xy - 2y^2 - 16x + 28y - 8 = 0$$

using MATLAB to plot the rotated and translated conic, the rotated conic, and then finally the conic.

7. We have defined $x \cdot y = x' * y$. Given a matrix A we can define $x \cdot_A y = x' * A * y$ and by matrix multiplication we get that

$$x \cdot_A y = y \cdot_A x$$
$$x \cdot_A (y + z) = x \cdot_A y + x \cdot_A z$$

When A is positive definite we get the additional properties

$$x \cdot_A x \geq 0$$
$$x \cdot_A x = 0 \iff x = \vec{0}$$

These four properties are the defining properties of a real inner product (a complex inner product uses $u \cdot v = \overline{v \cdot u}$ instead of $x \cdot y = y \cdot x$.) From here we can go on to define a norm $\|x\|_A = \sqrt{x' * A * x}$. This inner product will have its own opinions about orthogonality, $x \cdot_A y = 0$, and unit length, $\|x\|_A = 1$. Let $A = [1, 1; 1, 2]$ and first check that A is positive definite. Now write out by hand

$$[x \quad y] \begin{bmatrix} 1 & 1 \\ 1 & 2 \end{bmatrix} \begin{bmatrix} x \\ y \end{bmatrix}$$

Our first exercise is to plot the "unit circle" for this norm,

$$\{x \in \mathbb{R}^2 : \|x\|_A = 1\}$$

Get a unit circle in the Euclidean norm

```
t=0:pi/20:2*pi; x=cos(t); y=sin(t);
```
Now for each vector v=[x(i);y(i)] on the circle when we divide
```
v/sqrt(v'*A*v)
```

we get a vector of unit length in the **A-norm.** Do this for each point on the circle and plot the result.

To find an orthonormal basis for $\mathbb{R}^2$ with respect to this inner product, start with e_1, e_2 and perform the Gram-Schmidt procedure (see *Orthonormal Bases*) using the inner product

$x \cdot_A y$ in place of $x \cdot y$. This will produce two vectors u_1, u_2. Using `vecplot` (see page 43 of *Graphics*) plot u_1 and u_2 together. You may want to square the axes in the plot.

8. The best way to determine if a matrix is positive definite is not to find the eigenvalues but a variation of the LU Decomposition called the Cholesky Decomposition. This is represented in MATLAB by a function called `chol`. The matrix `R=chol(A)` is an upper triangular matrix with positive entries on the diagonal with the property that `R'*R=A` when `A` is positive definite. If `A` is not positive definite, `chol` returns an error message. See *The LU Decomposition* for more information. Try `eig(A)` on `hilb(5)`, `pascal(5)` and `A` defined by `B=list(5)`, `A=B'*B` to determine if they are positive definite. Now try `chol(A)` to determine if they are positive definite. Using `flops` (see *Flops*) determine which runs faster `chol` or `eig`.

COMMENTS

For more fun with conics see *Projective Geometry*.

The Singular Value Decomposition

ABSTRACT

The singular value factorization is able to convey very complete information about a matrix and its associated subspaces. The use of orthogonal matrices makes this factorization a highly accurate method of computation.

MATLAB COMMANDS

```
svd, norm, eig, cond, pinv, rank, null, plot
```

LINEAR ALGEBRA CONCEPTS

Singular Value Decomposition, Column Space, Null Space, Rank, Eigenvalues, Pseudo-inverse, Norm, Projection

BACKGROUND

We will jump right in with the statement of the singular value decomposition.

Theorem 1. *The Singular Value Decomposition. Given any $m \times n$ real matrix A there is an $m \times m$ orthogonal matrix U and an $n \times n$ orthogonal matrix V such that*

$$U^T A V = S = \begin{bmatrix} diag(\sigma_1, \ldots, \sigma_r) & 0 \\ 0 & 0 \end{bmatrix}$$

where $\sigma_1 \geq \sigma_2 \geq \ldots \geq \sigma_r > 0$.

First note that there are no complex matrices in the conclusion. The σ's are not eigenvalues of A. They are called the **singular values.** In the following discussion we will assume the notational convention

$$U = [u_1, \ldots, u_m] \qquad V = [v_1, \ldots, v_n]$$

and $AV = US$.

The Column Space

Look at $Av_1 = \sigma_1 u_1, \ldots, Av_r = \sigma_r u_r$. Thus $\sigma_1 u_1, \ldots, \sigma_r u_r$ are all in the Column Space of A. Since V is invertible, rank(A) = rank(AV) = rank(US) = r, since $\sigma_1 u_1, \ldots, \sigma_r u_r$ are orthogonal. Thus $u_1, \ldots, u_r$ form a basis for the Column Space of A. In MATLAB

```
A=magic(4)
[U,S,V]=svd(A)
```

Here $r = 3$ and so

```
B=U(:,1:3)
```

is an orthonormal basis for the Column Space of A. The remaining vectors $u_{r+1}, \ldots, u_m$ are a basis for the orthogonal complement of the Column Space of A, which is the Null Space of A^T (see *Projections.*)

The Null Space

Now $Av_j = \sigma_j u_j = \vec{0}$ for $j > r$ and so $v_{r+1}, \ldots, v_n$ are in the Null Space of A and since we know that the dimension of the Null Space is $n - r$, $v_{r+1}, \ldots, v_n$ is a basis for the Null Space of A. The vectors $v_1, \ldots, v_r$ are a basis for the orthogonal complement of the Null Space of A, which is the Column Space of A^T. In MATLAB

```
A*V(:,4)
```

The Eigenvalues of $A^T A$ and AA^T

Starting with $AV = US$ and taking the transpose we get $V^T A^T = S^T U^T$, using orthonormality, $A^T U = V S^T$. Looking at the ith column, $A^T u_i = \sigma_i v_i$. Now multiplying by A

$$AA^T u_i = A\sigma_i v_i = \sigma_i A v_i = \sigma_i^2 u_i$$

so that σ_i^2 is an eigenvalue of AA^T and u_i is an eigenvector. Similarly $A^T A v_i = \sigma_i^2 v_i$. Thus $\sigma_1^2, \ldots, \sigma_n^2$ are the eigenvalues of $A^T A$ and $\sigma_1^2, \ldots, \sigma_m^2$ are the eigenvalues of AA^T. In MATLAB

```
svd(A)
sqrt(eig(A'*A))
sqrt(eig(A*A'))
```

Norm and Condition Number

$\|A\|_2 = \|UAV\|_2 = \|S\|_2$, by Theorem 5 of *Orthonormal Bases*. By the definition of the 2-norm, $\|S\|_2 = \max_{\|x\|=1} \|Sx\|_2$. We want to maximize $f(x) = \sigma_1^2 x_1^2 + \cdots + \sigma_n^2 x_n^2$ subject to $x_1^2 + \cdots + x_n^2 = 1$. Taking partial derivatives $\frac{\partial f}{\partial x_i} = 2\sigma_i x_i$, and setting these to 0, subject to the constraint, shows that the solution must be $x = e_i$ for some i where e_i is the ith standard basis vector. But $Se_i = \sigma_i e_i$ and $\|\sigma_i e_i\|_2 = \sigma_i$ is maximum for $i = 1$. Thus $\|A\|_2 = \|S\|_2 = \sigma_1$.

If A is invertible, then $\sigma_r = \sigma_n \neq 0$ and as above $\|A^{-1}\|_2 = \|S^{-1}\|_2 = \frac{1}{s_n}$. The condition number is given by

$$\text{cond}_2(A) = \frac{\|A\|_2}{\|A^{-1}\|_2} = \frac{\sigma_1}{\sigma_n}$$

In MATLAB

```
cond(A)
S(1,1)/S(4,4)
```

MATLAB uses svd to compute cond in this manner.

Pseudo-inverse

When A is not invertible or even square, we can try to find an "inverse" by inverting the singular values. This process is the ***pseudo-inverse*** and we will write it in the MATLAB form pinv(A). First look at the $m \times n$ matrix

$$S = \begin{bmatrix} \text{diag}(\sigma_1, \ldots, \sigma_r) & 0 \\ 0 & 0 \end{bmatrix}$$

Now form the $n \times m$ matrix (Watch that size!)

$$\text{pinv}(S) = \begin{bmatrix} \text{diag}(\frac{1}{\sigma_1}, \ldots, \frac{1}{\sigma_r}) & 0 \\ 0 & 0 \end{bmatrix}$$

We do not attempt to find $\frac{1}{\sigma_i}$ when $\sigma_i = 0$, we simply leave it 0 in the pseudo-inverse. In MATLAB

```
P=pinv(S)
P*S
S*P
```

We now define pinv(A) for $U^T A V = S$ as

$$\text{pinv}(A) = V \text{ pinv}(S) \, U^T$$

Try

```
Q=V*P*U'
```

Do not expect great things to happen with this definition.

```
pinv(A)
Q*A
A*Q
```

Even though this pseudo-inverse is not an inverse, it can be shown that the pseudo-inverse has the following properties

    (1) $A \text{ pinv}(A) \, A = A$
    (2) $\text{pinv}(A) \, A \text{ pinv}(A) = \text{pinv}(A)$
    (3) $(A \text{ pinv}(A))^T = A \text{ pinv}(A)$
    (4) $(\text{pinv}(A) \, A)^T = \text{pinv}(A) \, A$

We can illustrate these in MATLAB

```
A*Q*A-A
Q*A*Q-A
(A*Q)'-A*Q
(Q*A)'-Q*A
```

Projections

Recall that the projection of b onto the Column Space of A is Ax where x is a solution to the normal equations $A^T A x = A^T b$, see Theorem 3 of *Projections*. Let $x = \text{pinv}(A) \, b$ and check using (3) and (1) above

$$(A^T A) \text{ pinv}(A) \, b = A^T (A \text{ pinv}(A))^T \, b = (A \text{ pinv}(A) \, A)^T b = A^T b$$

Thus $\text{pinv}(A) \, b$ is a solution to the normal equations and $A \text{ pinv}(A) \, b$ is the projection of b onto the Column Space of A. In *Least Squares* it is shown that a solution to the normal equations is a least squares solution to $Ax = b$ and so $x = \text{pinv}(A)\,b$ is a solution to the least squares problem.

Rank Approximations

Notice that when we take an $m \times 1$ vector, like u_i, and an $n \times 1$ vector, like v_i, and we form the outer product $u_i v_j^T$, we get a $m \times n$ matrix (see page 32.) With that in mind define the **rank k approximation to A**

$$A_k = \sigma_1 u_1 v_1^T + \cdots + \sigma_k u_k v_k^T$$

When we multiply

$$U^T A_k V = \begin{bmatrix} \text{diag}(\sigma_1, \ldots, \sigma_k) & 0 \\ 0 & 0 \end{bmatrix}$$

and we notice that $\text{rank}(A_k) = k$ when $k \leq r$. Look at $U^T(A - A_k)V = U^T A V - U^T A_k V$, the singular values $\sigma_1, \ldots, \sigma_k$ are subtracted out, leaving only $s_{k+1}, \ldots, s_r$. As above we compute the 2-norm

$$\|A - A_k\| = \|U^T(A - A_k)V\| = \sigma_{k+1}$$

when $k < r$. And when $k = r$, $A - A_r = 0$. The matrix A_k is the rank k approximation to A. With the rank r approximation being $A_r = A$. Try this in MATLAB

```
A=magic(4); [U,S,V]=svd(A);
A1=S(1,1)*U(:,1)*V(:,1)'
rank(A1), svd(A1)
A2=A1+S(2,2)*U(:,2)*V(:,2)'
rank(A2), svd(A2)
A3=A2+S(3,3)*U(:,3)*V(:,3)'
rank(A3), svd(A3)
A4=A3+S(4,4)*U(:,4)*V(:,4)'
rank(A4), svd(A4)
A4-A
```

PROBLEMS

1. Try MATLAB's `[U,S,V]=svd(A)` on the following matrices:

    (1) `rand(5)`
    (2) `list(5)`
    (3) `magic(8)`
    (4) `rand(5,1)*rand(1,8)`

In each case check that `U'*A*V-S` is appropriately small.

2. Use `svd` to compute the rank of each of the matrices in problem 1. You may want to use `format long` to get a better view of the singular values.

3. We can see the singular values and vectors in the 2-dimensional case. Form a unit circle $x^2 + y^2 = 1$ using `t=0:pi/25:2*pi; y=sin(t); x=cos(x);` Let `A=rand(2)` and plot how `A` distorts the circle by `z=A*[x;y]; x1=z(1,:); y1=z(2,:); plot(x1,y1)` Let `[U,S,V]=svd(A)`, hold the first plot and using `vecplot` from page 43 of *Graphics* plot `S(1,1)*U(:,1)` and `S(2,2)*U(:,2)`.

4. Use `svd` to compute the 2-norm and the condition number of the following matrices `vander(1:5)`, `hilb(7)`, and `magic(7)`. In each case check your computation with `cond(A)` and `norm(A)`. Another way to compute the 2-norm is to find `sqrt(`λ`)` where λ is the largest eigenvalue of `A'*A`. Try this on `A=hilb(7)`.

5. According to the Background discussion you can find orthonormal bases for the Column Space and the Null Space of A by selecting vectors from U and V. Find a basis for the Null Space and the Column Space for each of the following matrices: `hilb(7)`, `list(5)`, `magic(8)`, and `rand(5,1)*rand(1,8)`. Check you work by checking `A*x` for vectors which are supposed to be in the null space and checking that `A*x=u` has a solution for vectors u which are supposed to be in the column space.

6. Let `A=magic(8)` and form `pinv(A)`. Verify properties (1) – (4) of the pseudo-inverse using these values for `A` and `pinv(A)`. Now let `b=(1:8)'` and form the projection `p=A*pinv(A)*b`. Check that p is the projection by showing `p'*(b-p)=0`. Find a matrix A where $A\text{pinv}(A) \neq I_n$.

7. Compute the rank k approximation for

   `A=compan([1 1 1 1 0])`

for $k=1,\ldots,\text{rank(A)}$. Check that $\text{rank}(A_k)=k$ and $\| A-A_k \| = \sigma_{k+1}$. Check that for $k=\text{rank(A)}$, A_k-A is appropriately small.

8. Write MATLAB function `B=rankapp(A,k)` which returns the rank k approximation to `A`. Check your work against the example in problem 5.

9. You can find the singular value decomposition using `eig` by the following method which can also be used to prove Theorem 1. Let `A=rand(5,2)` and `[V,D]=eig(A'*A)`. We know that the rank of A is 3 so choose U as follows:

   `U=A*V`

   `U(:,1)=(1/norm(U(:,1)))*U(:,1)`

   `U(:,2)=(1/norm(U(:,2)))*U(:,2)`

   `U(:,3)=(1/norm(U(:,3)))*U(:,3)`

The columns of U were orthogonal, they are now orthonormal.

   `W=null(A')`

This gives us an orthonormal basis for the Null Space of `A'` which is the same as the orthogonal complement of the Column Space of `A`. Putting U and W together will give us an orthogonal matrix.

   `U=[U,W]`

   `U'*A*V`

   `svd(A)`

Try this for `A=list(5)` (see page 34 of *Building Matrices* for `list`.) You will need to rearrange some columns in U and V to get this to work properly.

10. This problem shows how a rank k approximation can be used to approximate data while significantly reducing the size of the stored data. Let A=mod(magic(26),2) (see page 16 for mod in the *MATLAB Tutorial*) and let B be the rank 4 approximation to A. What is the total number of entries in all the vectors and singular values used in the rank 4 approximation? You should get a number about $1/3$ the size of the 26×26 matrix which has 26^2 entries. A is a binary matrix while B is not, but we can convert it into a binary matrix easily with B=(B > .5); By checking s=sum(sum(abs(A-B))) you can learn the number of errors. What is the ratio of errors to the total number of entries in A? If you want to graphically view this try mesh(A), hold, mesh(B). You can see the differences while MATLAB is graphing.

11. The rank of a matrix is difficult to compute. Take the example of

    A=hilb(5)^3.

Check

    rank(A)

    rref(A)

Both agree that the rank is 4. Yet we know that hilb(5) is invertible and hence A is invertible with rank 5.

Rigid Frameworks

ABSTRACT
A framework is a simple bar and joint structure in either R^2 or R^3. We will analyze frameworks for the types of movements they can make, with particular emphasis on determining when a framework is rigid.

MATLAB COMMANDS
```
rank, null, *
```

LINEAR ALGEBRA CONCEPTS
Row Space, Null Space, Rank, Dimension

BACKGROUND

Motions

Consider a square formed by making joints at the ends of 4 bars of equal length. We will assume that the attachments at the joints are flexible so that the bars can rotate at the joints. Now place the square on a table and move it around on the table. Broadly speaking the following things can happen:

    (1) the whole square will be translated to a new position

    (2) the square will be rotated around an axis

These first two possibilities are called rigid motions since all the distances between the joints are being kept constant.

    (3) by holding one bar fixed the square will become a parallelogram and only distances along the bars will be preserved.

This last possibility we call a *finite motion.* By introducing a single bar along the diagonal of the square we can eliminate this finite motion and the result is a *finitely rigid* framework. A small wiggle in the framework could become a serious problem for a structural engineer. We will now address the study of *infinitesimal motions*. We will see that this framework has no nontrivial infinitesimal motions as well, but there are frameworks which are finitely rigid but do contain infinitesimal motions.

A *framework* is a set of joints $v_1, \ldots, v_n$ some of which are joined by bars of length r_{ij}. Suppose that there are m of these bars. For the time being we will assume that the joints are all points $v_i = (x_i, y_i)$ in R^2. To model the framework we will take the m equations which say that if the bar from v_i to v_j has length r_{ij}, then the distance from v_i to v_j is r_{ij}.

$$\sqrt{(x_i - x_j)^2 + (y_i - y_j)^2} = r_{ij}$$

Squaring gives

$$(x_i - x_j)^2 + (y_i - y_j)^2 = r_{ij}^2$$

Now suppose that each joint is given as a function of t so that $v_i(t) = (x_i(t), y_i(t))$, then differentiating with respect to t gives

$$2(x_i - x_j)(\alpha_i - \alpha_j) + 2(y_i - y_j)(\beta_i - \beta_j) = 0$$

where $\alpha_i = \frac{dx_i(t)}{dt}$ and $\beta_i = \frac{dy_i(t)}{dt}$, and α_j, β_j are the derivatives for (x_j, y_j). Rewriting this gives

$$(x_i - x_j)\alpha_i + (y_i - y_j)\beta_i + (x_j - x_i)\alpha_j + (y_j + y_i)\beta_j = 0$$

There are m of these equations; one for each bar. An assignment of nonzero velocities which satisfies this system of linear equations is called an **infinitesimal motion.** Let us return to the square, which we realize as

$$v_1 = (0,0) \quad v_2 = (1,0) \quad v_3 = (1,1) \quad v_4 = (0,1)$$

with bars connecting $v_1 v_2$, $v_2 v_3$, $v_3 v_4$, and $v_4 v_1$. We will call this framework S. The four equations become

$$(x_1 - x_2)\alpha_1 + (y_1 - y_2)\beta_1 + (x_2 - x_1)\alpha_2 + (y_2 - y_1)\beta_2 = 0$$
$$(x_2 - x_3)\alpha_2 + (y_2 - y_3)\beta_2 + (x_3 - x_2)\alpha_3 + (y_3 - y_4)\beta_3 = 0$$
$$(x_3 - x_4)\alpha_3 + (y_3 - y_4)\beta_3 + (x_4 - x_3)\alpha_4 + (y_4 - y_3)\beta_4 = 0$$
$$(x_1 - x_4)\alpha_1 + (y_1 - y_4)\beta_1 + (x_4 - x_1)\alpha_4 + (y_4 - y_1)\beta_4 = 0$$

Arranging these equations in matrix form yields a $m \times 2n$ matrix M

$$M = \begin{bmatrix} v_1 - v_2 & v_2 - v_1 & \vec{0} & \vec{0} \\ \vec{0} & v_2 - v_3 & v_3 - v_2 & \vec{0} \\ \vec{0} & \vec{0} & v_3 - v_4 & v_4 - v_3 \\ v_1 - v_4 & \vec{0} & \vec{0} & v_4 - v_1 \end{bmatrix}$$

with the variables $[\alpha_1, \beta_1, \alpha_2, \beta_2, \alpha_3, \beta_3, \alpha_4, \beta_4]^T$. The infinitesimal motions are the vectors in the Null Space of M. Let's put this in MATLAB

```
M=[-1,0,1,0,0,0,0,0
0,0,0,-1,0,1,0,0
0,0,0,0,1,0,-1,0
0,-1,0,0,0,0,0,1]
```

Now we can determine the dimension of the Null Space of M with

```
8-rank(M)
```

And we see that it is 4, which seems like a lot of motions but remember that we have the translations and rotations in there. In fact we can see them. In MATLAB let

```
hor=[1,0,1,0,1,0,1,0]
```

This motion is assigning a velocity of 1 to each joint in the positive x direction and 0 in the y direction. It thus represents a horizontal translation of the entire framework. We see quickly that hor is in the Null Space of M.

```
M*hor'
```

Similarly, the vertical translation is given by the vector

```
ver=[0,1,0,1,0,1,0,1]
```

and this is also in the Null Space of `M`.

    M*ver'

The rotations can be thought of as assigning a velocity to v_i in a direction perpendicular to the vector v_i. A vector perpendicular to a joint $v_i = (x_i, y_i)$ is $(y_i, -x_i)$. Look at the vector obtained by applying this perpendicular velocity at each joint.

    r=[0,0,0,-1,1,-1,1,0]

Again we see

    M*r'

We would like to remove these motions so that we can see what the "internal" motions of the framework will be. To eliminate translations and rotations we will simply include the vectors `r`, `ver`, and `hor` in the Row Space.

    A=[M;hor;ver;r]

Now we check the dimension of the Null Space of `A`

    8-rank(A)

We see that the dimension is 1. A way to view the process of adding rows is to let N be the Null Space of M, and W, the Row Space of M. The orthogonal complement of N is the Column Space of M^T which is the Row Space of M, namely W. We know from *Projections* that

$$n = \dim(W) + \dim(N)$$

so that changing M by adding elements of N to W should cause $\dim(W)$ to increase and $\dim(N)$ to be smaller. This has happened in this case. To see the motion try either of these. See page 122 of *The Null Space* for `nullbase`.

    null(A)

or

    nullbase(A)

A single vector has been returned, $[1, 1, 1, -1, -1, -1, -1, 1]^T$ (or a scalar multiple of this vector) which we can see has forces $(1, 1)$ on v_1, $(1, -1)$ on v_2, $(-1, -1)$ on v_3, and $(-1, 1)$ on v_4. This is a motion which pulls two of the diagonal joints together and pushes the other two apart. This happens to be a finite motion. Every finite motion is an infinitesimal motion. The Null Space of `A=[M;hor;ver;r]` is called the **Motion Space of S,** denoted $\mathcal{M}$.

Now suppose that we place a bar from v_1 to v_3. This means that we get a new equation

$$(x_1 - x_3)\alpha_1 + (y_1 - y_3)\beta_1 + (x_3 - x_1)\alpha_3 + (y_3 - y_1)\beta_3 = 0$$

Now include the vector which goes with this equation

    w=[-1,-1,0,0,1,1,0,0]
    M=[M;w]
    8-rank(M)

We see that $\dim(\mathcal{M}) = 0$ and there are no nontrivial infinitesimal motions. This new framework with the fifth bar is infinitesimally rigid. We say that S is ***infinitesimally rigid*** if $\mathcal{M} = \{\vec{0}\}$, equivalently, $\dim(\mathcal{M}) = 0$. For an example of a framework which has no finite motions yet is not infinitesimally rigid see Problem 1.

Forces

A *force* on S will be regarded as a vector $F = [F_1, \ldots, F_n] \in R^2$ where each F_i is a 2-dimensional vector representing the force applied to the joint v_i. Each element of the Row Space of M can be considered a force. A force in the Row Space of M is called a *resolvable force,* and the Row Space of M is called the **Space of Resolvable Forces,** denoted $\mathcal{R}$. Suppose that $F = [F_1, \ldots, F_n]$ is a resolvable force, so that there is a vector $s = [s_1, \ldots, s_m] \in R^m$ such that $F = sM$. If we look at the force F_j on the joint v_j we get

$$F_j = \sum_{i=1}^{m} s_i M(i, 2j - 1 : 2j)$$

that is $F_j - \sum_{i=1}^{m} s_i M(i, 2j - 1 : 2j) = \vec{0}$ so that the total force at each vertex is 0, hence "resolved."

The component F_i of a force $F = (F_1, \ldots, F_n)$ at the joint v_i can be given Plücker coordinates $(F_i, \det([F_i; v_i]))$. For $F = (\alpha, \beta)$ and $v = (x, y)$, $\det([F; v]) = [\alpha, \beta] \cdot [y, -x]$ indicates the size of the projection of F onto the perpendicular of v. We say that $F = (F_1, \ldots, F_n)$ is an **equilibrium force** if $\sum_{i=1}^{n} F_i = \vec{0}$ and $\sum_{i=1}^{n} \det([F_i; v_i]) = 0$. If we define $T : R^{2n} \to R^3$ by

$$T(F_1, \ldots, F_n) = (\sum F_i, \sum \det([F_i; v_i]))$$

then we get a linear transformation with matrix

$$E = \begin{bmatrix} 1 & 0 & 1 & 0 & \ldots & 1 & 0 \\ 0 & 1 & 0 & 1 & \ldots & 0 & 1 \\ y_1 & -x_1 & y_2 & -x_2 & \ldots & y_n & -x_n \end{bmatrix}$$

The Null Space of E is the **Space of Equilibrium Forces,** denoted $\mathcal{E}$. Notice that $E = [hor, ver, r]$, the vectors from the discussion above and $A = [M; E]$. The next series of results tell us how these spaces are related and will give us a simple criterion for infinitesimal rigidity.

Theorem 1. *With $\mathcal{R}$, $\mathcal{E}$, and $\mathcal{M}$ as defined above*

(1) $\mathcal{R} \subseteq \mathcal{E}$
(2) $\mathcal{E} = \mathcal{R} + \mathcal{M}$
(3) $\dim(\mathcal{E}) = \dim(\mathcal{R}) + \dim(\mathcal{M})$

PROOF:

(1) This is a minor abuse of notation since $\mathcal{R}$ is a set of row vectors and $\mathcal{E}$ is a set of column vectors. We will show that if u is a row of M, then $Eu^T = \vec{0}$, it then follows that $\mathcal{R} \subseteq \mathcal{E}$. Suppose that $u = [\cdots \; v_i - v_j \; \cdots \; v_j - v_i \; \ldots]$, where the dots indicate 0's. Then $E(1:2)u^T = v_i - v_j + v_j - v_i = \vec{0}$ and

$$\det\left(\begin{bmatrix} v_i - v_j \\ v_i \end{bmatrix}\right) + \det\left(\begin{bmatrix} v_j - v_i \\ v_j \end{bmatrix}\right) = \det\left(\begin{bmatrix} -v_j \\ v_i \end{bmatrix}\right) + \det\left(\begin{bmatrix} -v_i \\ v_j \end{bmatrix}\right)$$

$$= \det\left(\begin{bmatrix} -v_j \\ v_i \end{bmatrix}\right) - \det\left(\begin{bmatrix} -v_j \\ v_i \end{bmatrix}\right)$$

$$= 0$$

using the rules for applying elementary row operations and the determinant.

(2) Since $\mathcal{M}$ is the Null Space of $[M; E]$ and $\mathcal{E}$ is the Null Space of E, we have $\mathcal{M} \subseteq \mathcal{E}$, and so $\mathcal{R} + \mathcal{M} \subseteq \mathcal{E}$. Now given any $x \in \mathcal{E}$, since $\mathcal{R}^{\perp} = N$ where N is the Null Space of M, by Theorem 3 of *Projections* there is a $y \in \mathcal{R}$ and a $z \in N$ such that $x = y + z$. Now $y \in \mathcal{E}$ and $x \in \mathcal{E}$ and $z = x - y$, so $z \in \mathcal{E}$. Thus $\mathcal{E} = \mathcal{R} + \mathcal{M}$.

(3) By Theorem 8 of *Dimension and Rank* we get

$$\dim(\mathcal{E}) = \dim(\mathcal{R} + \mathcal{M}) = \dim(\mathcal{R}) + \dim(\mathcal{M}) - \dim(\mathcal{R} \cap \mathcal{M})$$

If $x \in \mathcal{R} \cap \mathcal{M}$, then $x \cdot x = 0$ and so $x = \vec{0}$. Thus $\mathcal{R} \cap \mathcal{M} = \{\vec{0}\}$ and $\dim(\mathcal{R} \cap \mathcal{M}) = 0$. ∎

Theorem 2. *If $n \geq 2$, then $\dim(\mathcal{E}) = 2n - 3$.*

PROOF: Suppose that v_1 and v_2 are two distinct joints, then look at the following piece of E:

$$\begin{bmatrix} 1 & 0 & 1 & 0 \\ 0 & 1 & 0 & 1 \\ y_1 & -x_1 & y_2 & -x_2 \end{bmatrix}$$

Since either $y_1 \neq y_2$ or $x_1 \neq x_2$, the rank$(E) = 3$, and hence $\dim(\mathcal{E}) = 2n - 3$. ∎

This last theorem gives an easy method for checking rigidity. The advantage of this theorem is that we do not need to deal with the vectors hor, ver, and r. The next theorem allows us to settle the rigidity question for planar frameworks quickly.

Theorem 3. *If $n \geq 2$, the following are equivalent:*
   (1) *S is infinitesimally rigid*
   (2) *$\mathcal{E} = \mathcal{R}$*
   (3) *rank$(M) = 2n - 3$.*

PROOF: We will show (1) → (2) → (3) → (1).
(1) → (2). If $\mathcal{M} = \{\vec{0}\}$, then by Theorem 1(2), so $\mathcal{E} = \mathcal{R}$.

(2) → (3). If $\mathcal{E} = \mathcal{R}$, then by Theorem 1(3),

$$\text{rank}(M) = \dim(\mathcal{R}) = \dim(\mathcal{E}) = 2n - 3$$

by Theorem 2.

(3) → (1). If $\text{rank}(M) = 2n - 3$, then by Theorem 1(3), $\dim(\mathcal{M}) = \dim(\mathcal{E}) - \dim(\mathcal{R})$ and so $\dim(\mathcal{M}) = (2n - 3) - (2n - 3) = 0$ by Theorem 2. ∎

Pinning Joints

Now suppose we want to play with a framework by only fixing some of the joints while letting others be moved by translations and rotations. Return to the original matrix for the square

$$M = \begin{bmatrix} -1 & 0 & 1 & 0 & 0 & 0 & 0 & 0 \\ 0 & 0 & 0 & -1 & 0 & 1 & 0 & 0 \\ 0 & 0 & 0 & 0 & 1 & 0 & -1 & 0 \\ 0 & -1 & 0 & 0 & 0 & 0 & 0 & 1 \end{bmatrix}$$

Suppose that we want to pin down $v_1 = (0, 0)$. This is essentially the same as removing the translations at v_1. Adjoin the vectors

$$[1, 0, 0, 0, 0, 0, 0, 0] \qquad [0, 1, 0, 0, 0, 0, 0, 0]$$

On doing this we see that the dimension of the Null Space is 2. The rotation is still present as well as constrained translations of the other joints. What happens when you pin v_1 and v_2? What happens when you pin v_1 and v_3?

Three Dimensional Frameworks

To work in three dimensions we follow essentially the same approach. Given a framework with n joints, $v_1, \ldots, v_n$ where $v_i = (x_i, y_i, z_i)$, and m bars, we create the matrix M with m rows, one for each bar, and $3n$ columns, the columns of $M(:, 3i - 2 : 3i)$ are devoted to vertex v_i. If there is a bar from v_i to v_j, then for some k, $M(k, 3i - 2 : 3i) = v_i - v_j$ and $M(k, 3j - 2 : 3j) = v_j - v_i$. A framework which is rigid in two dimensions need not be rigid in three dimensions and in most cases will not be. There are additional rigid motions in $\mathbb{R}^3$. In $\mathbb{R}^3$ there are three independent directions to make translations and three independent axes of rotation. This means that the rigid motions will have dimension 6. To remove the translations we adjoin the vectors

$$[e_1, e_1, \ldots, e_1] \quad [e_2, e_2, \ldots, e_2] \quad [e_3, e_3, \ldots, e_3]$$

where $e_1 = (1, 0, 0)$, $e_2 = (0, 1, 0)$ and $e_3 = (0, 0, 1)$. To remove a rotation around the x−axis, adjoin

$$[0, z_1, -y_1, 0, z_2, -y_2, \ldots, 0, z_n, -y_n]$$

around the y−axis adjoin

$$[z_1, 0, -x_1, z_2, 0, -x_2, \ldots, z_n, 0, -z_n]$$

and around the $z-$axis adjoin

$$[y_1, -x_1, 0, y_2, -x_2, 0, \dots, y_n, -x_n, 0]$$

A single force vector F acting on a joint v is given six dimensional *Plücker coordinates* by $[v - F, v \times F]$ where $v \times F$ is the vector cross product (see *Linear Transformations*). Each coordinate of $v \times F$ represents the moment of the force about a coordinate axis. Given a force $F = (F_1, \dots, F_n)$, there is a linear transformation $T : R^{3n} \to R^6$ defined by

$$T(F_1, \dots, F_n) = (\sum_{i=1}^{n} F_i, \sum_{i=1}^{n} (v_i \times F_i))$$

We say that F is an *equilibrium force* if $T(F_1, \dots, F_n) = \vec{0}$. The kernel of T is the *Space of Equilibrium Forces,* denoted $\mathcal{E}$. With minor modifications there are results parallel to Theorems 1, 2, and 3. We will only summarize without proof the main result.

Theorem 4. *If S is a three dimensional framework with $n \geq 3$, then S is infinitesimally rigid if and only if $\text{rank}(M) = 3n - 6$.*

PROBLEMS

1. Determine if the triangle given below is infinitesimally rigid.

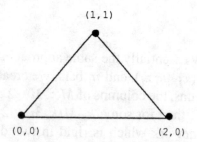

The degenerate triangle is given by $v_1 = (0, 0)$, $v_2 = (1, 0)$, and $v_3 = (2, 0)$ with bars from $v_1 v_2$, $v_1 v_3$, and $v_2 v_3$. Find an infinitesimal motion in the degenerate triangle. Is this a finite motion?

2. A five-point star can be made by forming the vertices

```
x=(1:5)';
v=[cos(2*pi*x/5),sin(2*pi*x/5)];
```

Let v_k =v(k,:). Then the five point star is given in the next figure. What is $\dim(\mathcal{M})$?

Add a bar from v_1 to v_2, is this framework infinitesimally rigid? Is there a finite motion? How many more bars need to be added to make this rigid?

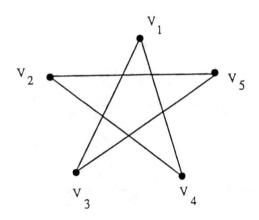

3. Consider the framework given below. If you can only place bars from top to bottom joints, where do bars need to be placed to make this rigid? In view of Theorem 3(2) we see that every equilibrium force is resolvable. Look at the force $F = (F_1, F_2, F_3, F_4, F_5, F_6)$ where $F_1 = (-7, -6)$, $F_2 = (-2, -15)$, $F_3 = (22, -24)$, $F_4 = (-18, 13)$, $F_5 = (-7, 14)$, $F_6 = (12, 18)$. Show that F is an equilibrium force on S. Find a resolution s such that $sM = F$.

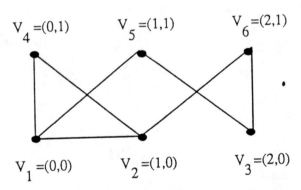

4. We are given a framework $v_1 = (0, 0)$, $v_2 = (1, 0)$, and $v_3 = (2, 0)$ with bars $v_1 v_2$ and $v_2 v_3$. In view of problem 1, we cannot make this rigid by merely adding bars. How would you add another joint and bars to get an infinitesimally rigid framework?

5. The following simple planar truss structure is proposed for a bridge. What is your analysis? Is it safe? Is there a finite motion of this structure?

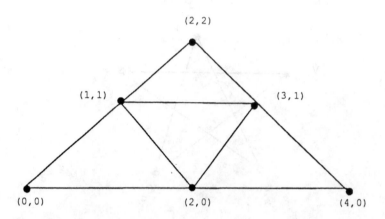

6. The following frameworks look similar. Are they both rigid? Is there a finite motion?

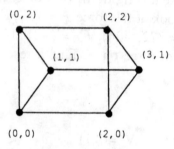 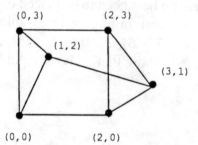

7. A realization of Kempe's Translator is given below. It is fixed at the origin, but the other joints can move freely in the plane. It is useful as a drafting aid, since a pencil attached at point Q moves in translation to the movement of P. Suppose that we fix P, is the result infinitesimally rigid? Is there a finite motion? Now release P and fix Q, is the result infinitesimally rigid? Is there a finite motion?

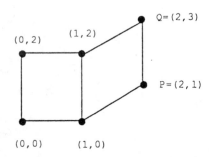

8. A realization of Sylvester's Pantograph is given below. It is fixed at the origin, but the other joints can move freely. This pantograph will both translate and scale. Suppose that we fix P, is the result infinitesimally rigid? Is there a finite motion? Now release P and fix Q, is the result infinitesimally rigid? Is there a finite motion?

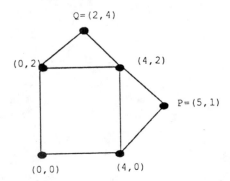

9. In three dimensions construct a tetrahedron on the joints $v_1 = (0,0,0)$, $v_2 = (1,0,0)$, $v_3 = (0,1,0)$ and $v_4 = (0,1,1)$ by placing a bar between each pair of joints. Is the tetrahedron infinitesimally rigid?

10. In three dimensions construct a cube on the joints $v_1 = (0,0,0)$, $v_2 = (1,0,0)$, $v_3 = (1,1,0)$, $v_4 = (0,1,0)$, $v_5 = (0,0,1)$, $v_6 = (1,0,1)$, $v_7 = (1,1,1)$, $v_8 = (0,1,1)$, What is the dim($\mathcal{M}$)? Can you account for all of these motions? Add bars v_1v_7, v_2v_8, and v_4v_6. Is the result infinitesimally rigid? Start over and add bars v_1v_6, v_1v_3, and v_1v_8. Is the result infinitesimally rigid?

MATLAB Index

This index provides references to MATLAB functions and programs in the context that they are used in this book. This index is not a comprehensive referencing of all MATLAB functions. The MATLAB `help` command will provide the current descriptions of MATLAB functions. There is a short list of MATLAB functions on page 24. The index of mathematical terms is on page 239.

Linear Algebra Index

This index provides references for those standard mathematical terms from linear algbra and numerical analysis which are used in this book. There is MATLAB Index on page 237 to help locate MATLAB functions and programs.